D0857922

Biochemical Basis of Plant Breeding

Volume II
Nitrogen Metabolism

Editor

Carlos A. Neyra, Ph.D.
Associate Professor of Plant Physiology
Biochemistry and Microbiology Department
Rutgers University
New Brunswick, New Jersey

CRC Press, Inc.
Boca Raton, Florida

Library of Congress Cataloging in Publication Data
Main entry under title:

Biochemical basis of plant breeding.

   Includes bibliographies and index.
   Contents: v. 1. Carbon metabolism — v. 2. Nitrogen
metabolism.
   1. Plant breeding—Collected works.   2. Botanical
chemistry—Collected works.   3. Plants—Metabolism—
Collected works.   I. Neyra, Carlos A., 1940- .
SB123.B5 1986   631.5'3   85-5760
ISBN-0-8493-5741-1 (v. 1)
ISBN-0-8493-5742-X (v. 2)

   This book represents information obtained from authentic and highly regarded sources. Reprinted material is quoted with permission, and sources are indicated. A wide variety of references are listed. Every reasonable effort has been made to give reliable data and information, but the author and the publisher cannot assume responsibility for the validity of all materials or for the consequences of their use.

   All rights reserved. This book, or any parts thereof, may not be reproduced in any form without written consent from the publisher.

   Direct all inquiries to CRC Press, Inc., 2000 Corporate Blvd., N.W., Boca Raton, Florida, 33431.

© 1986 by CRC Press, Inc.

International Standard Book Number 0-8493-5742-X

Library of Congress Card Number 85-5760
Printed in the United States

PREFACE

This book presents a comprehensive survey of progress and current knowledge of those biochemical processes with greater potential for the development of superior plant genotypes: photosynthesis, photorespiration, starch biosynthesis, nitrate assimilation, biological nitrogen fixation, and protein synthesis. The first three of these topics were elegantly covered in Volume I (Carbon Nutrition). The other topics are the components of Volume II entitled Nitrogen Nutrition. Throughout the contents, we have attempted to provide sufficient background information and discussion of biochemical tools to strengthen the approach and goals of plant breeding programs. Modern plant breeding, in the context of this book, is seen as a combination of both the traditional or conventional plant breeding approach with the more novel molecular and cellularly based genetic engineering methodologies. We anticipate that this new era of research will rely strongly on the progress made in plant biochemistry research and its applicability to the goal of improving the genetic potential of crops.

The significant growth of agricultural productivity over the past few decades can be attributed in large part to the application of modern concepts of genetics to the breeding of superior crops. The development of semidwarf wheat and rice genotypes together with the development of hybrid corn are clear examples of successful plant breeding aimed at increasing crop productivity. Nonetheless, the yield increases attained in all of those three cases were highly dependent on the availability and responsiveness to fertilizer nitrogen. Economic considerations and protection of environmental quality however, will eventually limit the amounts of fertilizer nitrogen to be used in farming. To maintain or even increase yields, genotypes will have to be developed with a greater efficiency in the use of soil or atmospheric nitrogen sources. In the case of cereal grains crops, genotypic differences in nitrogen utilization have been demonstrated, indicating that the potential exist for developing nitrogen-efficient genotypes (see Chapters 4 and 5). On the other hand, grain legumes are able to use both NO_3^- and N_2 as nitrogen sources but increasing N input for increased grain yields will place a larger demand on photosynthate to satisfy the energy required by the simultaneous or complementary reduction of NO_3^- (nitrate assimilation) or atmospheric N (N_2-fixation). Thus, lugume breeding programs should consider the increase in photosynthetic productivity simultaneously with the N-incorporation by these crops (see Chapters 2 and 6).

Nitrate is the predominant form of soil N available to crop plants under nonpaddy conditions and the primary step in the assimilation of NO_3^- is catalyzed by the enzyme nitrate reductase (NR). On the other hand, biological nitrogen fixation contributes with about 10×10^7 tons of nitrogen fixed per annum in agricultural soils. The primary step in the assimilation of atmospheric N is catalyzed by the enzyme nitrogenase. The properties and regulation of these two enzymes, nitrate reductase and nitrogenase, are thoroughly discussed in Chapters 1 and 2, respectively. The use of NR activity as a selection trait is discussed in Chapter 5 and the regulation of nitrate uptake, translocation, and reduction is dealt with in Chapter 4. A broad understanding of how genes code for proteins in nuclei, plastids and mitochondria, and the regulation of gene expression in plants is discussed in Chapter 3, devoted to protein synthesis.

Because of the vast knowledge needed to make an in-depth presentation of the subjects under consideration, I chose to invite some of the best regarded scientists to contribute their expertise to this book. Within broad editorial guidelines, the contributors have been responsible for the precise content and approach of their chapters. I express my sincere thanks to each of the authors for their response and excellent contributions to this book.

Carlos A. Neyra, Ph.D.
Editor

MAY 1 3 1987

THE EDITOR

Carlos A. Neyra, Ph.D., is an Associate Professor of Plant Physiology in the George H. Cook College of Agriculture and Environmental Sciences at Rutgers University, New Brunswick, New Jersey. Dr. Neyra received his B.S. degree in Agronomy in 1965 from the Universidad Nacional de Tucuman in Argentina and his Ph.D. degree in Plant Physiology in 1974 at the University of Illinois at Urbana-Champaign.

In addition to being a prolific writer in the fields of Agronomy, Agricultural Biochemistry, and Plant Physiology, Dr. Neyra is widely known internationally for his activities as a teacher, researcher, and Scientific Advisor in several countries including Argentina, Brazil, The Dominican Republic, Panama, Peru, and the U.S.

Dr. Neyra is a tenured member of the Rutgers faculty with full-member status in the following graduate programs: Botany and Plant Physiology, Horticulture, Microbiology, and Soils and Crops. He teaches formally both undergraduate and graduate courses of Plant Physiology and has served on numerous University committees. He is a reviewer for a number of refereed journals including Plant Physiology, Canadian Journal of Botany, Canadian Journal of Microbiology, Soil Sciences, and others. He has also collaborated as a peer reviewer for various granting agencies (USDA, NAS, NSF, AID, BARC, BOSTID, etc.). Dr. Neyra was elected a full-member of the honorary societies of Sigma Xi and Gamma Sigma Delta in 1974.

CONTRIBUTORS

Frederick E. Below
Research Associate
Department of Agronomy
University of Illinois
Urbana, Illinois

Robert H. Burris, Ph.D.
Professor of Biochemistry
Department of Biochemistry
University of Wisconsin
Madison, Wisconsin

Wilbur H. Campbell, Ph.D.
Associate Professor of Chemistry
Department of Chemistry
SUNY College of Environmental
  Science and Forestry
Syracuse, New York

R. T. Dunand
Research Physiologist
Louisiana State University
Rice Research Station
Crowley, Louisiana

Richard H. Hageman, Ph.D.
Professor of Crop Physiology
Department of Agronomy
University of Illinois
Urbana, Illinois

William A. Jackson, Ph.D.
Professor of Soil Science
Department of Soil Science
North Carolina State University
Raleigh, North Carolina

Eugene J. Kamprath, Ph.D.
William Neal Reynolds Professor of
  Soil Science
Department of Soil Science
North Carolina State University
Raleigh, North Carolina

R. J. Lambert, Ph.D.
Professor of Plant Genetics
Department of Agronomy
University of Illinois
Urbana, Illinois

Paul W. Ludden, Ph.D.
Assistant Professor of Biochemistry
Department of Biochemistry
University of Wisconsin
Madison, Wisconsin

Mark J. Messmer
Plant Breeder
Garst Seed Company
Ames, Iowa

Robert H. Moll, Ph.D.
Professor of Genetics
Department of Genetics
North Carolina State University
Raleigh, North Carolina

William L. Pan, Ph.D.
Assistant Soil Scientist
Assistant Professor of Soils
Department of Agronomy and Soils
Washington State University
Pullman, Washington

Carl A. Price, Ph.D.
Professor
Waksman Institute
Rutgers University
Piscataway, New Jersey

Joe H. Sherrard
Assistant Lecturer
School of Agronomy and Forestry
University of Melbourne
Parkville, Victoria, Australia

John Smarrelli, Jr., Ph.D.
Research Associate
Department of Biology
University of Virginia
Charlottesville, Virginia

Mark R. Willman
Research Assistant
Department of Agronomy
University of Illinois
Urbana, Illinois

Colleen Winkels
Research Assistant
Department of Agronomy
University of Illinois
Urbana, Illinois

TABLE OF CONTENTS

Chapter 1

NITRATE REDUCTASE: BIOCHEMISTRY AND REGULATION

Wilbur H. Campbell and John Smarrelli, Jr.

TABLE OF CONTENTS

I. BIOCHEMISTRY OF NITRATE REDUCTASE

A. Introduction and Scope

Nitrate reductase (NR) is the biocatalyst for the reduction of nitrate to nitrite. In eukaroytic organisms, this initial step in nitrate assimilation is driven by reduced pyridine nucleotides (NADH and NADPH). NR can now be recognized as a multicenter redox enzyme, because it contains several redox centers and becomes reduced during enzyme turnover. The redox centers have been shown to be flavin adenine dinucleotide (FAD), cytochrome b_{557}, and molybdenum. A Mo-cofactor has also been identified. The physiological activity of NR appears to occur by pyridine nucleotides donating electrons to the FAD to reduce NR, internal transfer of electrons from FAD to Mo via the cyt. b, and reduction of nitrate by the enzyme-bound reduced Mo. Apparent non-physiological activities of NR can also be demonstrated and will be called the dehydrogenase activity and the reductase activity. The dehydrogenase activity involves pyridine nucleotide electron donation to the FAD and reduction of one and two electron acceptors, such as cyt. c, ferricyanide, and dichlorophenol indophenol. These reductions may involve different redox centers of the enzyme. The reductase activity is less defined but electrons are donated to either the cyt. b or the Mo by reduced flavins or viologen dyes leading to the reduction of nitrate by the Mo center. These concepts are embodied in Figure 1. The fundamental principle underlying the catalytic mechanism of NR can be described as indirect transfer of electrons and protons from electron donors to electron acceptors via direct participation of the enzyme as an intermediary.

Since it is likely that NR is composed of a single polypeptide chain, it is a multifunctional protein. That is to say, it is likely to be composed of distinct domains where the polypeptide chain folds into more compact and somewhat separated subparts which have close contact. These domains are yet to be definitely demonstrated; however, sufficient evidence has accumulated to permit speculative models to be constructed. One possible model is discussed below in relation to Figure 2, where domains are assigned for the various prosthetic groups. While these two models serve to define the catalytic and structural properties of NR, they also offer a framework for the discussion of regulation of the enzyme. Operationally, there are two levels of regulation to be considered: the amount of the enzyme and its level of activity. The presentation here will focus on biochemical considerations of enzyme control and exclude detailed discussions of genetic control.

In terms of the overall scope of this review, we will limit our discussions to higher plant forms of NR and will only occasionally draw in information from algal and fungal systems. A variety of reviews of NR and nitrate reduction recently have been published and there appears no need to attempt to review all this material again.[1-7]

B. Types and Distribution of Nitrate Reductase

Two types of NR have been isolated and characterized from higher plant tissues. Evans and Nason[8] showed the presence of an NAD(P)H:NR in soybean leaves, which has been further studied more recently.[9] NAD(P)H:NR (EC 1.6.6.2) is bispecific for

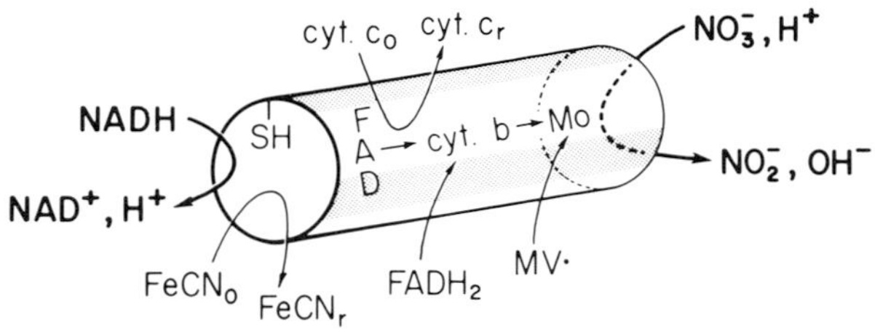

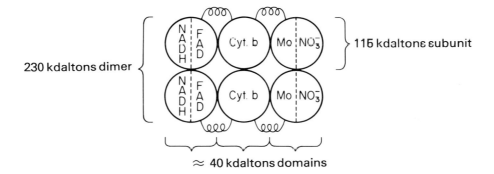

FIGURE 1. A schematic model of the catalytic properties of nitrate reductase. The physiological reaction of NADH and nitrate is shown in boldface. Partial reactions are shown in light type. NR is represented as a "mini-electron-transport" protein with one active site for NADH and a second for reducing nitrate. The active sites are joined by the electron transport chain of FAD, cyt. b, and Mo. A critical thiol group (-SH) is shown at or near the binding site for NADH. The dehydrogenase partial reaction is represented by showing the reduction of ferricyanide on the NADH side of FAD and the reduction of cyt. c on the other side of the FAD. The reductase partial activity is represented by the addition of electrons from $FADH_2$ to the cyt. b of the NR and from the reduced methyl viologen ($MV \cdot$) to the Mo. The exact physical sites for the interactions of these electron donors and acceptors is not known.

FIGURE 2. A schematic model of the structure of nitrate reductase. The structure of the native homodimer of squash cotyledon nitrate reductase is shown as deduced from recent data.[46 48,74] As determined by polyacrylamide gel electrophoresis, the molecular weight of the homodimer squash NADH:NR is 230 kilodaltons (KD) and the largest molecular weight found by SDS denaturation and electrophoresis is 115 kilodaltons. Three domains are shown, one for binding each of the NR prosthetic groups. A binding site for NADH is shown on the FAD domain, while the nitrate binding site is shown on Mo domain. It is assumed that both sets of active sites would be fuctional. The number and molecular weights of the domains are speculative, but based on sizes of the breakdown products observed after denaturation. The number of hinge regions in the polypeptide backbone, which are shown by the helical line joining hetero-domains in each monomer, is also unknown at this time.

pyridine nucleotide; that is to say, it will accept electrons from either NADH or NADPH. This enzyme form has been found in a number of tissues and will be described in more detail below. However, the most common form of higher plant NR is monospecific for NADH and currently is the most widely studied form of the enzyme.

The most completely studied examples of NADH:NR (EC 1.6.6.1) have been isolated from the tissues of corn, barley, wheat, spinach, squash, and tobacco.[10-15] A third type of higher plant NR may have been recently described in studies of "constitutive" or "noninducible" NR activity, soybean, and dry beans.[16,17,17a] Little is known about this form of NR and it may be shown eventually to have properties distinctly different from those of other higher plant forms of the enzyme.

With respect to the distribution of NR, the leaves of most plants are the site of the bulk of nitrate reduction but the roots of all plants probably contribute some nitrate reducing activity, especially to meet local needs.[18,19] Pate's[20] now classic analysis of bleeding sap showed that plants are basically of two types: those with little or no reduction of nitrate in the roots and those with almost all activity in the roots. While a great deal of study of nitrate reduction in plants has operated on the assumption that root reduction of nitrate was of little consequence, more recent studies have given a greater consideration to root nitrate reducing activity.[21-23] The roots of crop plants appear to contribute a small, but probably significant, amount to the total plant nitrate reduction.[24] This may be especially true in the seedling stage of growth after stored reserves have been exhausted, and during the period of growth when root tissue composes a high percent of the metabolically active tissue of the whole plant.

The NADH:NR is found in the leaves of all plants analyzed except *Erythrina senegalensis* which is a tropical legume with only NAD(P)H:NR thus far demonstrated.[25,26] The leaf cells of C_4 plants have differential localization of NADH:NR, where it is found only in mesophyll cells and not in bundle sheath cells.[27] NADH:NR has also been found in the storage organs and roots of plants.[8,18,19,25,28,28a] While NAD(P)H:NR is found in the leaves of some plants, it is nearly always accompanied by an NADH:NR and it is less widely distributed.[25] Most interestingly, for corn, which lacks any strong evidence for the presence of an NAD(P)H:NR in the leaves, this enzyme form has been found in scutella and roots.[25,28] In barley leaves, an NAD(P)H:NR has been found in plants which are mutants with respect to the wild-type NADH:NR.[29] It was recently suggested that NAD(P)H:NR may be a genetic possibility in all plants and that in most leaves this potential is suppressed by light, which would explain the more common occurrence in seedlings and nongreen tissues.[30] Monospecific NADPH:NR (EC 1.6.6.3) is found only in fungi and is not known to occur in higher plants or algae.[31] The most common form in algae is NADH:NR, which is similar in some respects to the leaf form of NADH:NR.[1] The most universal form is the bispecific NAD(P)H:NR which is found in fungi, algae, and higher plants. Immunochemical data indicate that all these forms of NR share a common ancestor and these data will be discussed in more detail below.[32]

C. Nitrate Reductase Assays

There are three types of assays for nitrate reductase activity: (1) whole plant assays; (2) intact tissue assays; and (3) in vitro assays.[33] Each of these assays measures a different aspect of nitrate reduction activity and that makes it difficult to compare the results achieved with them. The whole plant assay, which constitutes a true in vivo assay, is done by keeping a careful account of all nitrate taken up and reduced, and is often coupled to measured increases in total plant reduced nitrogen. Huffaker and co-workers have used this assay.[23,34,35] The intact tissue assay, which is often, but erroneously, referred to as an "in vivo" assay, measures the production of nitrite by tissue slices or pieces either in the presence or absence of nitrate.[33] This assay is widely used and exists in many modified forms.[36,37] It is rather dependent on the presence of surfactants or organic solvents and has been used as a measure of nitrate availability when done in the absence of nitrate.[37,38]

The in vitro assay is done on isolated preparations of NR and probably most directly measures the total capacity for the reduction of nitrate, when neither electron donor nor nitrate is limiting.[33] The selection of buffers and pH for these assays has focused on phosphate at pH 7.5, but some exceptions have been found and a proper assay procedure should involve optimization of these conditions as well as assuring that substrates are saturating. Since most of these assays are dependent on the diazotization of nitrite, it is important to insure that nothing is interfering with color development, which has been found to be a problem with excess NADH. Several methods have been found to overcome this problem.[39,40] Assays have also been described for the evaluation of NR partial activities, which must be done in vitro.[41,42] Also only the in vitro assay can be used to evaluate the electron donor specificity (i.e., monospecific for NADH or bispecific for NAD(P)H).[43,44] In demonstrating the utilization of NADPH as electron donor, it is important to show that the nucleotide is not hydrolyzed to NADH by the action of a phosphatase. Three different methods have been described to address the action of phosphatase on NADPH.[25,43,45]

D. Nitrate Reductase Preparations

The application of affinity chromatography, using blue-dye ligands based on Cibacron blue FG3-A, has become so wide that nearly all preparations of the enzyme are done with these tools.[10-15] Since these gels are of such wide interest, we have described their synthesis and use in Appendix A. While partial purification of higher plant NR is easily achieved with affinity chromatography,[41] the preparation of homogeneous enzyme has still proved difficult.[46] While homogeneous NR can be defined as preparations containing only NR as shown by gel electrophoresis using sensitive protein and enzyme-activity stains, it has been found that the specific activity of these homogeneous preparations varies.[46] Consequently, additional criteria may be needed to define NR preparations that contain intact and fully active enzyme. Among these measures of homogeneity we might cite the following: approach to a limiting value of specific activity such as 100 to 200 units/mg protein (1 unit = 1 μmol nitrite formed per minute), low ratios of partial activities to complete activity, evidence for a single polypeptide chain after denaturation and electrophoresis, and ratios of prosthetic group contents approaching reproducible and integral values. The goal of obtaining preparations of higher plant NR which meet these standards has not yet been achieved.

Highly purified NR has been prepared from the leaves of barley, corn, soybean, squash, spinach, tobacco, and wheat (Table 1). Soybean is a special case with both NADH:NR and NAD(P)H:NR in leaves and will be discussed below.[9,43] For barley, corn, tobacco, and wheat, the preparations are mainly of blue-dye affinity chromatography, purified enzyme and have specific activities of about 1 to 23 units/mg protein.[12-15] For spinach, affinity chromatography was combined with other methods to yield enzyme with 25 units/mg proteins.[10] By using blue Sepharose, squash NR can be prepared with specific activities from 2 to 19 units/mg protein depending on how strong the NR activity is in the starting tissue and how carefully the gel is washed after the enzyme is bound.[11,41,46] The combination of blue Sepharose and pink Sepharose has led to preparations with specific activities of 30 to 40 units/mg protein.[47,48] However, the combination of zinc chelate columns with blue Sepharose provides squash NADH:NR with specific activities of 110 units/mg protein.[46] The synthesis and use of these gels are described in Appendix A.

Part of the difficulty encountered in the preparation of higher plant NR is the instability of the enzyme. Various attempts have been made to stabilize the enzyme in the extracts of leaves or other tissues by the addition of proteins or other agents.[33] The inclusion of protectant against phenolics is well advised, as is the addition of EDTA to protect against heavy metal inhibition.[33,41,46,49-51] In some tissues such as the leaves of barley and corn, the addition of a thiol reductant like cysteine or dithiothreitol is re-

Table 1

SOME PROPERTIES OF NITRATE REDUCTASE

Plant or organism	Type	Specific activity (units/mg)	pH Optima	Mol wt[a] (kdaltons) Holo-NR	Apo-NR	Quaternary structure[b]	Ref.
Higher Plants							
Barley	NADH	8	7.5	200—220	100—110	A₂	15, 56
Corn	NADH	15	7.5	270	?	?	25, 60, 61
Soybean	NADH	0.4	6.5	330	?	?	9, 43
	NAD(P)H	2	6.5	220	?	?	9, 43
Squash	NADH	110	7.5	230	115	A₂	46, 48, 74
Spinach	NADH	24	7.5	200—240	120	A₂	1, 10, 48
Tobacco	NADH	1.2	7.5	200	110	A₂	14
Wheat	NADH	23	7.5	500	?	?	1, 12, 62
Algae							
Chlorella vulgaris	NADH	93	7.6	360	90	(A₂)₂	1, 63
Ankistrodesmus braunii	NAD(P)H	80	7.5	460	60	(AB)₄	1, 64
Fungi							
Neurospora crassa	NADPH	125	7.5	230	115	A₂	1, 65
Penicillium chrysogenum	NADPH	225	7.2	200	100	A₂	1, 66
Rhodotorula glutinis	NAD(P)H	148	7.5	230	118	A₂	1, 67

[a] Subunit molecular weight was either deduced by addition of cleavage fragment sizes or is represented by the size of the largest fragment.

[b] A₂ = homodimer; (A₂)₂ = dimer of homodimer or tetramer; and (AB)₄ = tetramer of heterodimer or octamer.

quired, while in others like spinach and squash no cysteine need be used.[46, 50-53] However, FAD appears to be a universal agent for stabilization of NR during preparation.[1,33,46,51] For protection against proteolytic degradation and possibly other beneficial effects, proteins such as bovine serum albumin or casein can be included in the extraction buffer,[13,28,33,51,54] but proteinase inhibitors have been used with leupeptin being the most effective in barley leaf extractions.[55,56]

E. Molecular Weight of Holo-Enzyme and its Subunit

Since NR is more asymmetric in molecular shape than most proteins, estimates of the molecular weight (M_r) by a single method such as gel filtration or sedimentation in a gradient will be in error.[1,57] By combining the Stokes radius from gel filtration and the sedimentation coefficient from gradient centrifugation, the M_r of the native enzyme can be calculated.[58] Alternatively, the M_r can be estimated by gel electrophoresis using gels of different percent acrylamide to yield different mobilities for the enzyme.[59] The degree of change of the electrophoretic mobility vs. the gel porosity (percent acrylamide) appears to yield the native M_r when appropriate standards are used.[47,48] The native M_r of NR are shown in Table 1. The electrophoretic method tends to yield higher values for the native M_r than the combined method.[1,48,60,61] Sedimentation equilibrium yielded the highest estimate.[62] However, it appears that the native M_r for higher plant NADH:NR is from 200 to 240 kilodaltons (kdaltons).

The subunit M_r for NR is not so clearly established and has only recently been as heavily investigated as the native M_r.[1,15,47,48,56,68-70] The results to be considered here are of two types: those where partially purified enzyme is centrifuged in a sucrose gradient and those where homogeneous NR is denatured and electrophoresed.[47,48,56,68-72] The latter method provides the most definitive information and has been done by different procedures by different groups.[48,56,68] These data are summarized in Table 1. Except for one study with barley leaf,[56] all studies yield multiple subunits with large differences in M_r, with the lower limit being 37 to 45 kdaltons and an upper limit of 100 to 120 kdaltons.[1,14,48] Wray and co-workers[55,73] recently compared the subunit sizes found for barley leaf NR preparations, which had been made with and without leupeptin to protect the enzyme against proteolytic degradation. They found that more bands were observed and greater amounts of low M_r bands when leupeptin and other protectants were omitted.[15,73] They concluded for barley leaf that the lower M_r bands resulted from degradation of an intact subunit of M_r equal to 100 kdaltons.[15] The results of Warner, Kleinhofs, and co-workers[13,56] agree with Wray's in finding a 100-kdalton subunit for barley NR, but differ in that lower M_r polypeptides were shown to be unrelated to the NR subunit with respect to Cleveland peptide maps. The studies done of squash cotyledon NADH:NR have led us to conclude that the intact subunit of this enzyme is a 115-kdalton polypeptide.[74] In the case of squash, we originally found evidence for two subunits with M_r of 44 and 75 kdaltons,[47,48,70] but subsequent studies done with higher specific activity enzyme preparations have demonstrated the presence of the 115-kdalton subunit and a number of lower M_r bands.[74] In view of the evidence for specific enzymes capable of cleaving the NR molecule being in plant tissues and the high lability of NR in crude extracts,[75] it might be expected that NR would be isolated in a partially degraded form. Although the addition of a protease inhibitor to the buffers used in preparation of *Neurospora crassa* NADPH:NR provided stabilization of this enzyme,[76] in general, these problems with enzyme degradation have not been found in the studies of algal and fungal NR.[1,31] In most cases, the subunit M_r could be attributed to a single band or doublet of 90 to 130 kdaltons (Table 1).

F. Nitrate Reductase Prosthetic Groups

The identification of the prosthetic groups of NR as FAD, cyt. b_{557}, Mo, and Mo-cofactor has been accomplished but no quantitative data have been presented.[1] The flavin of higher plant NR appears to be tightly held, although the addition of FAD to assays tends to stimulate the activity of some plant NR.[1] FAD is clearly useful in stabilizing NR during purification or heat denaturation.[1,46] In experiments recently done with squash NADH:NR, enzyme bound to pink Sepharose could be partly eluted by FAD, while the bulk of the enzyme was eluted by NADH, but the amount of NR eluted did not depend on the order of elution.[48] These results tend to indicate that deflavo-forms of NR were present in these preparations, and these deflavo-forms might account for the stimulation of activity by FAD. The cytochrome component of NR was first identified by Garrett and Nason[77] with *N. crassa* NADPH:NR and has since been identified as a component of spinach, barley, tobacco, and squash NADH:NR.[1,10,14,50,78] Hewitt and co-workers[79] found the redox potential of the cyt. b of spinach NR to be -60 mV. They have also shown that [99]Mo is bound into spinach NR and that an inactive tungsten analogue may be formed in plants in the absence of Mo.[80,81] They have also done extensive studies of a reputed Mo-cofactor;[1] however, more recent work with other Mo-containing enzymes has indicated that the Mo-cofactor is not a protein component but is more likely to be a low M_r organic molecule related to urothione, which is a pterin.[82-85] The presence of this moiety was recently reported in higher plants and the study of this substance as a component of higher plant NR will probably be done in the near future.[86-89]

G. Nitrate Reductase Structural Models

Two different structure models must be considered for higher plant NADH:NR. One model is a dimer of two subunit monomers, which has been presented by Hewitt and Notton.[1] The other model is the homodimeric one first presented by Pan and Nason[65] for *N. crassa* NADPH:NR but modified to contain equal amounts of all prosthetic groups. The Hewitt model assumes that NR is composed of two types of subunits of about equal M_r which are each present twice, and when combined with a low M_r Mo-cofactor yields the native enzyme. This model is somewhat consistent with the observed subunit sizes for NR, but fails to realistically deal with the high likelihood that purified preparations of NR consist of a mixture of cleaved and intact forms of the enzyme. Wray and co-workers[15,55,68,69,71-73] have embraced the latter explanation and offer the idea that the lower M_r NR fragments represent the domains of the enzyme which are resistant to proteolyltic cleavage. They envision labile bonds between the stable domain structures of the enzyme, which are cleaved during purification. The concept of NR with domains and labile bonds connecting the stable substructures is presented in Figure 2. In order to simplify this model, the native and subunit M_r for squash NADH:NR has been used in this model, such that the native NR is composed of two subunits of 115 kdaltons to yield a holoenzyme of 230 kdaltons.[46,48,70,74] These values would differ slightly if either barley or spinach data were used.[1,15,56] The domain size is based on the observations made with denatured squash NR and correlates well with the often observed 40 KD cyt. c reductase activity of higher plants.[68,71,72,90] The molecular size of the domain must be considered tentative at this time and, in addition, the number of domains is a matter of speculation. We suggest there may be three domains: (1) for binding FAD, (2) for binding the cyt. b, and (3) for binding the Mo and Mo-cofactor. The flavin domain must also contain the binding site for pyridine nucleotides to donate electrons to the FAD. The nitrate binding site must be either wholly or partially on the Mo/Mo-cofactor domain. Phosphate probably binds to the Mo/Mo-cofactor domain. Both the NADH and FAD binding sites are expected to be composed of β-sheet secondary structures as have been described for the glutathione reductase tertiary structure based on X-ray diffraction data.[91] The domain binding the heme-iron of cyt. b may be predominantly α-helical structure like has been found for cyt. b_5 or cyt. b_{562}.[92]

For the Mo-containing domain, no proteins containing Mo have been crystallized and so no suggestion can be made for the expected secondary structure.

H. Nitrate Reductase Catalytic Properties

The reduction of nitrate by reduced pyridine nucleotides has a large negative delta G of 34 kcal/mol and formally would be considered a redox reaction. The reduction of NR by pyridine nucleotides can be shown by observing either the visible spectrum or electron paramagnetic resonance spectrum.[1,50,76] The reduced NR can be reoxidized by the addition of nitrate. In combination with the knowledge of the prosthetic group content of the enzyme, it can be concluded that NR is a multicenter redox enzyme.[1,93] This class of enzymes includes xanthine oxidase, glutamate synthase, nitrogenase, hydrogenase, glutathione reductase, nitrite reductase, ferredoxin:$NADP^+$ reductase, and others.[91,93,94] A subclass of these multicenter redox enzymes can be defined as the Mo-containing enzymes, of which NR is obviously a member and includes xanthine oxidase, aldehyde oxidase, xanthine dehydrogenase, sulfite oxidase, nitrogenase, and formate dehydrogenase.[1,93-95] The properties that these enzymes share, in addition to having Mo as a prosthetic group, are the presence of multiple redox centers including flavin, heme-iron, or iron-sulfur. Most commonly their native structures are composed of two identical subunits (i.e., homodimeric) and they have well-defined domains as substructures of their subunit. Since some of these enzymes have been studied in much greater detail than NR, many observations made about these multicenter redox enzymes can be useful in the study of NR.

1. Reactions Catalyzed by Nitrate Reductase

The reduction of nitrate by reduced pyridine nucleotides as catalyzed by NR can be represented by two half reactions:

$$NADH + NR \text{ (oxidized)} \rightarrow NAD^+ + H^+ + NR \text{ (reduced)}$$

$$NO_3^- + H^+ + NR \text{ (reduced)} \rightarrow NO_2^- + OH^- + NR \text{ (oxidized)}$$

Net reaction:

$$NADH + NO_3^- \rightarrow NAD^+ + NO_2^- + OH^-$$

The overall reaction and each half-reaction are essentially irreversible.[1,66] The first half-reaction appears to involve the FAD portion of NR and can be linked to the dehydrogenase partial activity of NR, which is dependent on FAD.[1] The second half-reaction appears to involve the Mo-containing portion of NR and can be linked to the reductase partial activity of NR, which is dependent on the presence of Mo.[1] The species called "NR (reduced)" in these half-reaction equations should be taken to mean a two-electron reduced form where the electrons may be found in any of the redox centers. While the redox centers are often represented as a linear sequence with unidirectional flow of electrons from FAD to Mo (see Figure 1), it is reasonable to consider the redox centers in rough equilibrium and that the direction of electron flow will depend on what electron acceptor is present.[76,93] In other words, the internal flow of electrons is not irreversible and electrons that reside in the Mo center may flow back to the FAD if the only electron acceptor present reacts only with the FAD site of the enzyme. For xanthine oxidase, the redox potential of its prosthetic groups may depend on the presence or absence of products, which indicates the need for detailed study of electron transfer properties of multicenter redox enzymes.[93-95] Also as has been found

Table 2

RELATIVE RATES OF REACTIONS CATALYZED BY SQUASH
COTYLEDON NADH:NR[a]

Reaction type	Electron donor	Electron acceptor	Relative rate
Complete	NADH	Nitrate	100
Dehydrogenase	NADH	Ferricyanide	970
		Cytochrome c	340[b]
		Dichlorophenolindophenol	250
		Menadione	440
		Methylene blue	820
		Benzoquinone	820
		2,3-dichloro-1,4-naphthoquinone	270
Reductase	FADH$_2$	Nitrate	40
	Methyl viologen	Nitrate	210

[a]   Blue Sepharose purified NR with a specific activity of 3.2 units/mg protein.
[b]   This rate is lower (140) in more highly purified enzyme, while the methyl viologen
       reductase activity ratio remains the same (210).

for xanthine oxidase, NR may accept electrons at redox centers other than the FAD
and donate electrons from centers other than the Mo.

2. Alternate Substrates for Nitrate Reductase

Since NR has the kinetic properties of an enzyme with two separate nonoverlapping
and nonidentical active sites, which will be explained fully below, alternate substrates
for the electron donor and acceptor sites can be considered separately.[96] The pyridine
nucleotides, NADH and NADPH, are the only electron donors for the FAD site of
NR (Figure 1). The site of electron donation by reduced FAD or viologen dyes is not
established with certainty, but these nitrate reducing activities do not depend on a
functional flavin and appear to involve other redox centers.[1] However, chlorate and
bromate appear to compete with nitrate and to be alternate substrates for the electron
donating site of NR.[1] We have also studied the iodate reducing activity of squash
NADH:NR and found that NADH was oxidized rapidly by the enzyme in the presence
of iodate.[97] However, iodate did not disappear to any significant degree. With chlorate, the reaction product released from NR is chlorite, but with bromate and iodate as
substrate the corresponding two-electron reduced species (i.e., bromite and iodite) are
considered metastable and probably disproportionate. Thus, the six electron reduction
attributed to *E. coli* NR in catalyzing the reduction of iodate to iodine,[98] probably
occurs via a combination of enzymatic reduction and chemical reactions between the
unstable reduction products and iodate at or near the active site. Since the product
chlorite is toxic to both NR and living organisms, the formation of chlorate-resistant
mutants is a highly convenient test for selecting mutants impaired in the ability to use
nitrate for growth.[99]

3. Partial Reactions of Nitrate Reductase

Several catalytic activities copurify with NR and may be classed as two different
types: the dehydrogenase and the reductase activities.[1] These reactions are shown sche-
matically in Figure 1 and comparative reaction rates for the complete and partial activities of squash NADH:NR are given in Table 2.[41,46] Both wheat and squash NADH:NR
have been shown capable of catalyzing the NADH-dependent reduction of quinones.[41,100] With the exception of the reduction of iron chelates such as ferric citrate,
which will be discussed below, the partial activities are considered to be artificial activ-

ities of NR and to have no physiological significance. The importance of the partial activities lies in the insight into the catalytic properties of NR provided by their study. Both the complete and the partial reactions have been shown to have different thermal inactivation rates and the partial reactions have been found to be differentially denatured.[12,101] Both the complete NR and dehydrogenase partial activity are inhibited by agents which react with enzyme sulfhydryls such as *p*-hydroxymercuribenzoate.[1,102,103] These agents when added at low concentrations do not affect the reductase activity and the enzyme can be protected against inhibition by the presence of pyridine nucleotides.[102,103] Metal binding agents like cyanide and azide inhibit both the complete NR and the reductase partial activities without detectable inhibition of the dehydrogenase activity.[1,8] Inhibition with heavy metals like cupric ion or by hydroxylamine results in loss of the complete NR activity without affecting either partial activity.[50,53,104] In summary, the partial reactions permit easy detection of the selective effects of inhibitors of NR and differentiation among the operational characteristics of the enzyme. These characteristics can be defined in a general sense as an electron donor site which requires a thiol group, a functional electron transfer chain, and a substrate reduction site.

4. Iron-Chelate Reduction by Nitrate Reductase

The reduction of ferricyanide by pyridine nucleotides as catalyzed by NR may represent an in vivo activity, which is better explained in terms of a general capability of NR to catalyze the reduction of iron-chelates. Over the last few years, the iron assimilation pathway of higher plants has been characterized.[105] Iron is taken up from the soil by excreted agents and enters the root as reduced iron where it is oxidized. Oxidized iron is transported to the leaves as ferric citrate. However, only ferrous iron is used by ferrochetalase, the enzyme responsible for incorporation of iron into porphyrins and other metabolic products, so that ferric citrate must be reduced before the iron can be used by the leaf. No agent has been characterized as the iron-citrate reductase of the leaf. However, it was shown recently that squash NADH:NR could reduce ferric citrate at rates of about 5% of nitrate reduction, but with a pH optimum of 6.5.[74] It also seems likely that the low M_r cyt. c reductase activities extracted from higher plants may also be involved in iron reduction.[90] The possibility that NR is involved in iron reduction is supported by the findings that iron assimilation is hampered by nitrate, cupric ions, and foliar phosphates, which would all influence the ability of NR to act as an iron reductase.[105] While other enzymes in chloroplasts and mitochrondria may also act as iron-chelate reductases, the soluble phase enzyme responsible for this acitvity in higher plant leaves may be NR.

5. Steady-State Kinetic Mechanism of Nitrate Reductase

Based on initial velocity analysis of spinach NADH:NR, the kinetic mechanism of the enzyme was described as ping-pong or substituted enzyme type.[1] This suggestion would account for the participation of oxidized and reduced NR, and the two half-reactions shown before. However, when the study of higher plant NADH:NR kinetics was extended to include product inhibition analyses, it was found that these patterns were not consistent with a classical ping-pong mechanism but could be accommodated by the two-site ping-pong mechanism.[11,52,96] This mechanism is very appealing because it explains not only the involvment of reduced NR and the half reactions, but also the separation of the electron donor site and nitrate reduction site.[11] The two-site ping-pong mechanism for NR suggests that two physically separated, nonoverlapping active sites (one for the electron donor and a second for the reduction of nitrate) are found on the enzyme and that the two sites are connected by electron transfer via the enzyme's redox centers (FAD, cyt. b, and Mo). This type of steady-state kinetic mechanism has been suggested for other Mo-containing enzymes and for multicenter redox enzymes in general.[93,94] More recently it has been suggested for a fungal NADPH:NR and a detailed derivation of a rate equation was presented.[66]

While the two-site ping-pong kinetic mechanism appears to account adequately for the observed kinetic properties of NR, the initial velocity kinetics of *Chlorella* NADH:NR were recently shown to be of the intersecting type which would not fit a ping-pong type of mechanism.[106] It was suggested that this NR uses a type of random sequential mechanism involving two sites. These intersecting line initial velocity patterns have also been observed for fungal NR.[107,108] In addition, another kinetic mechanism involving an isomerization of the enzyme was described for spinach NADH:NR and an algal NAD(P)H:NR.[109,110] These diverse results tend to indicate that the kinetic mechanism for NR is yet to be defined and that further study is required. While we feel that further study would be beneficial to gain a better understanding of NR kinetics, it is also possible to reconcile all the existing results in terms of the original suggestion we made for explaining the kinetics of higher plant NADH:NR.[11,96]

We first suggested that NR kinetics could follow one of several patterns for addition of substrates and departure of products. One possible pathway was the classic ping-pong with NADH reducing NR and NAD^+ leaving before nitrate was bound and reduced.[11] However, since NR is a two-site enzyme, both NADH and nitrate could be bound simultaneously. This latter pathway involves a ternary complex between the substrates and the NR, which would yield intersecting line reciprocal plots in the initial velocity studies unless catalytic events following the reduction of the enzyme were rate limiting.[11,106] Thus, the initial velocity pattern for NR is more an indication of the importance of binary and ternary complexes in the rate equation rather than of different kinetic mechanisms for the enzyme. Palmer and Olson[93] have offered a general kinetic mechanism for multicenter redox enzymes which incorporates the concept that multisite enzymes can utilize either binary or ternary complexes. This complexity is illustrated by the kinetic studies of xanthine dehydrogenase where parallel line initial velocity kinetics were observed at higher concentrations of NAD^+ but intersecting patterns at low concentrations of NAD^+.[111] Thus, it appears that conceptually both the parallel line initial velocity patterns generally observed for higher plant NR and the intersecting patterns observed for some other NR can be accommodated by the two-site ping-pong mechanism, which we originally proposed for corn and squash NADH:NR.[11,96] A more detailed discussion of the NR kinetic mechanism has been presented in regard to recent studies of the kinetics of soybean leaf NAD(P)H:NR.[103,112]

I. Catalytic Model for Nitrate Reductase

The turnover numbers and kinetic constants for several NR from higher plants and other sources are compared in Table 3. As can be seen from this table, a great deal of similarity is found among the kinetic properties of these forms of NR. So while these enzyme forms differ in structural properties with some being described as homodimers and others as tetramers or octamers, they all seem similar in kinetic properties and on the level of a single subunit to be similar in chemical properties such as prosthetic group ratios.[1] Thus, all forms of NR seem to be described by a catalytic model like the one shown in Figure 1. However, the influence quaternary structure (i.e., dimeric, tetrameric, and octameric form) has on catalysis has not, as yet, been established. However, it has been reported that the dimer and tetrameric forms of *C. vulgaris* NADH:NR had the same specific activity.[63] While squash and other plant NR appear to be a mixture of dimer and tetramer forms, it is clear that most of the enzyme exists as a dimer.[46,48,70] However, the balance or equilibrium between dimer and tetramer has not been studied, especially in relation to specific activity.

Finally, consideration should be given to the implications of the two-site character

Table 3
KINETIC CONSTANTS FOR NITRATE REDUCTASE[a]

Enzyme source	k_{cat} (s^{-1})	Km$_A$	Km$_B$	k_{cat}/Km$_A$	k_{cat}/Km$_B$	Ref.
		$\times 10^6 M$		$\times 10^{-7} s^{-1} M^{-1}$		
Higher Plants						
Curcurbita maxima[b]	210	5	40	4.2	0.53	11, 46
Spinacea oleracea[b]	48	5	50	1.0	0.1	1, 10
Algae						
Ankistrodesmus braunii[c]	160	23	150	0.7	0.1	110
Chlorella vulgaris[b]	140	4	30	3.5	0.47	106
Fungi						
Neurospora crassa[c]	240	62	200	0.4	0.1	1
Penicillium chrysogenum[c]	375 (140)[d]	10 (6)	91 (23)	3.8 (2.3)	0.41 (0.61)	66, 108
Rhodotorula glutinis[c]	290	20	125	1.5	0.23	67

[a] Using the data given in Table 1, k_{cat} = (specific activity) × (subunit molecular weight) × (1/60 s^{-1}).
[b] Substrate A = NADH and B = nitrate.
[c] Substrate A = NADPH and B = nitrate.
[d] Data in brackets are for pH 6.2.

of NR. As discussed above in relation to the kinetic mechanism, NR has two active sites: one for pyridine nucleotides to donate electrons to the enzyme and one for the transfer of electrons to nitrate. This two-site arrangement makes it possible for nitrate to bind to the enzyme (possibly in a rapid equilibrium) and to wait for the electrons to arrive for its reduction. It would appear from the kinetic studies that nitrate binding to NR would have little impact on the reduction of the enzyme by NADH.[11,96,103,112] Alternatively, the enzyme may also be able to accept more than one pair of electrons from NADH without obligatory reduction of nitrate.[76] However, reduction of nitrate may occur with a higher rate from a preferred reduction state of the enzyme.[93] In fact, the substrate inhibition usually observed for NADH might be attributed to "over-reduction" of NR leading to a slower capability for reduction of nitrate.[47,113] In any case, the possibility of accepting more than the required two electrons (i.e., electron storage capacity) and the presence of separate active sites for the electron donor and acceptor, tend to give NR and related multicenter redox enzymes a flexible kinetic response.[76,93] Thus, NR catalytic activity may depend on the relative ratio of substrates as well as their absolute concentrations. For *E. coli* glutamate synthase, this flexibility of kinetic response was described as making the enzyme very sensitive to the redox state of the cell and as a possible regulatory control of the enzyme, which would not depend on any change in the chemical or physical structure of the enzyme.[114] This type of responsive enzyme might be described as "self-regulatory" and this may be an important consideration in the case of NR, for which no feedback inhibitors or other well-defined mechanisms of regulation have been found.

J. NAD(P)H:Nitrate Reductases

Originally the only NR found in soybean leaves was a bispecific NAD(P)H:NR; however, it was subsequently shown that the NADH:NR of this tissue was more labile than the NAD(P)H:NR.[8,115] The discovery of the thiol-dependent NADH:NR in soybean leaves led to the suggestion that the activity with NADPH might be due to a phosphatase that actively converted NADPH to NADH and that NADH:NR was the only form of this enzyme in higher plants.[49,116] However, Jolly[117] was able to separate two forms of NR from the extracts of soybean on DEAE-cellulose. It was demon-

Table 4

PROPERTIES OF HIGHER PLANT NAD(P)H: NITRATE REDUCTASES

Source	pH Optima	Activity ratio NADPH:NADH	Km (mM)				
			NO$_3^-$				
			NADPH	NADH	NADPH	NADH	Ref.
Soybean leaf	6.5	1.1	7.5	6.7	0.002	0.003	9, 103
Soybean leaf[a]		1.0	9.5	16.5	0.0003	0.002	9, 103
Corn root	7.5	1.2	0.3	0.2	ND	ND	25
Corn scutellum	7.5	1.7	0.6	ND	ND	ND	28
Rice seedling	7.4	1.6	1.1	0.4	ND	ND	45
Barley leaf	7.7	4.5	1.2	1.5	0.17	0.11	29
Erythrina leaf	7.5	1.1	10	8	0.014	0.013	26

[a]   The values shown here were measured at pH 7.5.

strated that an NAD(P)H:NR form was present in these extracts and that phosphatase activity could not account for these results.[43] In the last few years, NAD(P)H:NR has been found in the tissues of five species of higher plants and some of the properties of these enzyme forms are summarized in Table 4. In addition to being bispecific for pyridine nucleotide, these forms of NR have unusually high Km values for nitrate when compared to the higher plant NADH:NR (see Table 3). Soybean is even more unusual because it has a pH optimum of 6.5, which is a full pH unit lower than other higher plant NR and the lowest reported for any NR.[8,9,43,117] Since soybean has such unusual catalytic properties, the kinetic mechanism might also be unusual. The analysis of the kinetic mechanism was done at both pH 6.5 and 7.5 with NADH and NADPH as electron donor.[103,112] Parallel line reciprocal plots were found for all four sets of conditions which, when combined with the appropriate product inhibition studies, indicated that soybean NAD(P)H:NR functions with a two-site ping-pong kinetic mechanism at both pHs with either electron donor.[103] In addition, it was shown that NADH and NADPH donate electrons to the same reaction site on this bispecific NR.[103,112] Therefore, despite the unusual properties of soybean NAD(P)H:NR, it appears to have the same kinetic mechanism as other higher plant NR.[11]

In terms of the physiological significance of NAD(P)H:NR forms, several different cases must be examined. First, in soybean and corn tissues, the NAD(P)H:NR is always accompanied by an NADH:NR, but these two NR forms appear to differ in kinetic response to nitrate concentrations such that the bispecific form would be more effective at higher concentrations of nitrate.[8,25,112] This differential response to nitrate for the two forms of NR seems to be most highly refined in soybean where the NADH:NR is inhibited at high nitrate where NAD(P)H:NR is most active.[8,103] In addition, in soybean cotyledons the NADH:NR reaches maximum activity at a younger age than the NAD(P)H:NR.[44] For rice seedlings, the two enzymes forms can be differentially induced; NAD(P)H:NR was induced by phytohormones and nitrophenyl compounds, but less strongly by nitrate, which mainly induced the NADH:NR.[45] In barley NADH:NR mutants like nar la, the gene for expression of NAD(P)H:NR appears to be derepressed as an adaption to the loss of the dominant NADH:NR, which is the only NR form found in the wild-type plants.[29,113] For *E. senegalensis,* the only form of NR present is the bispecific NAD(P)H:NR, which is unusually active.[26] It is worth noting that the other legumes such as *Phaseolus vulgaris* and *Vicia faba* appear to contain only NADH:NR and that the NAD(P)H:NR is not universally present in this family of plants.[1,49] However, the genetic capability for the synthesis of NAD(P)H:NR may be present in all plant tissues and regulated by override mechanisms to prevent the induction of this enzyme form in the presence of nitrate. One suggestion is that light may suppress the derepression of the NAD(P)H:NR gene,[30] but it has not been demonstrated that all nongreen tissues contain NAD(P)H:NR. However, the NAD(P)H:NR forms of higher plant have not been widely studied and in general appear to be less stable than the NADH:NR forms and may be more difficult to detect.[25]

K. Immunochemistry of Nitrate Reductase

The difficulty found in trying to prepare homogeneous higher plant NR has delayed the preparation of monospecific antisera and the application of immunochemical methods for NR. Hewitt and co-workers[118,119] prepared anti-NR against partially purified spinach enzyme and used the anti-NR to detect apo-NR. Their assay was based on the ability of the anti-NR to inhibit the NADH:NR activity of a chosen preparation of purified enzyme. With this assay they found that Mo-deficient plants with very low NR activity had about 30% of the apo-NR found in control nitrate-grown plants.[119] An earlier study of wild type and mutant fungi had also used anti-NR in an inhibition of activity immunoassay, where it was suggested that this method could only detect the complexed form of NR and not its subunits.[120] Amy and Garrett[121] did a detailed study of NR immunochemistry in the wild type and mutants of *N. crassa*. Their data tend to indicate that the monospecific anti-NR could detect not only the active enzyme but also its subunits using both inhibition of activity and rocket immunoelectrophoresis. Funkhouser and Ramadoss[122] have prepared monospecific anti-NR against homogeneous *C. vulgaris* NADH:NR and found evidence for the presence of apo-NR in ammonia-grown cells using rocket immunoelectrophoresis. Anti-NR prepared against barley NADH:NR have been used to assay for cross-reacting material (CRM) in mutants and wild type plants.[123] Based on these results, the mutants can be broken into three groups for the apparent structural or nia mutants: (1) mutants Az 12, Az 29, and Az 30 had essentially no CRM; (2) Az 13, Az 23, Az 31, and Az 33 had intermediate amounts of CRM; and (3) Az 28 and Az 32 had higher than control levels of CRM.[123] For Mo-cofactor or cnx mutants, the level of CRM was intermediate. It is interesting to note that the mutants with the lowest levels of CRM contain a moderately active NAD(P)H:NR.[29]

Monospecific anti-NR has been prepared with the homogeneous enzyme from squash cotyledons.[32,53,124] Using this anti-NR in immunoprecipitation and inhibition of activity assays, the cross reactivities of seven forms of NR were compared.[32] With squash NADH:NR as the standard, spinach NADH:NR was found to be nearly identical, while corn NADH:NR was closely related but not as close as spinach. Soybean NAD(P)H:NR and NADH:NR were both distantly related to squash as were *C. vulgaris* NADH:NR and *N. crassa* NADPH:NR. More recently, this antiserum against squash NR was shown to have high cross reactivity with the partially purified NADH:NR from tobacco and barley.[125] In order to develop a quantitative immunochemical assay for detection of NR in crude extracts, an enzyme-linked immunosorbent assay or ELISA has been developed (Figure 3).[53,124] This is a sandwich type assay and does not require homogeneous NR as a competitive assay would. This assay can detect NR protein in the range of 1 to 10 ng, which is in the range of the amounts of active NR found in crude extracts. These low levels of NR are difficult to detect in crude extracts with methods such as immunoelectrophoresis which depend on the formation of an immunoprecipitate.[124] This assay not only detects holo-NR and heat-denatured NR, but also the fragments of NR present in crude extracts that do not bind to blue Sepharose.[53,124] The utility of this ELISA method for the analysis of NR biosynthesis in squash cotyledons is under current investigation.[124] A preliminary report has appeared describing similar investigations of NR biosynthesis in barley leaves using anti-NR and rocket immunoelectrophoresis.[126]

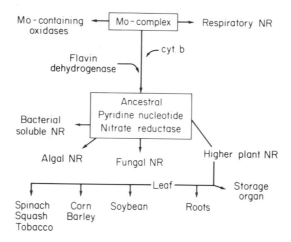

FIGURE 3.   An immunologically based family tree for nitrate reductase. Relationships among various forms of NR and related enzymes are represented as based on the degree of immunological cross reactivity among the isolated proteins and various forms of anti-NR.[1,2,32,121-125] We speculate that the Mo-complex is the most primitive component of NR, since NR shares the same Mo-cofactor with other Mo-containing enzymes. Respiratory NR appears to contain this same Mo-cofactor. However, to evolve into a pyridine nucleotide using enzyme, the Mo-complex must have been combined with a cyt. b component and a flavin dehydrogenase. If these components are housed in separate domains, the genetic material coding for these domains may also be separated into exons in the NR gene as has been found for the globin genes. Therefore, the comparisons illustrated here based on immunological studies will be better delineated once the NR gene is isolated from a higher plant and has been sequenced.

While biochemical and physical properties of NR indicate that these similar enzymes probably form a group related in genetic origin or a homologous family, the recent immunochemical data strongly support this view.[32,53,125] The evidence for the homology of the NR enzymes may be enhanced by the refinement of immunochemical studies, such as may result from the preparation of monoclonal antibodies aganist NR. More definitive proof of structural similarities among the NR forms will be found when the amino acid sequences of these proteins or the genes coding for them have been elucidated and can be compared. In Figure 4, we offer a speculative family tree for NR. Here we have drawn together data on both biochemical and immunochemical properties to deduce relationships among NR and related enzymes. The feature that most unites these proteins is the Mo-cofactor and the protein which binds it. While we indicate here that the bispecific forms of NR may be more closely related to their corresponding monospecific forms, it could be that the NAD(P)H:NR forms will form a separate group. In fact, the data on barley mutants tend to indicate that the NAD(P)H:NR form is not very cross reactive with the antibody against the wild-type NADH:NR.[123]

Clearly, the immunochemical study of NR has opened new possibilities for the study of this enzyme. It can be expected that the immunochemical methods will be very useful in the study of NR biosynthesis and turnover.[53,126] Furthermore, it is likely that monospecific anti-NR will also be useful in the immunoprecipitation of NR which are

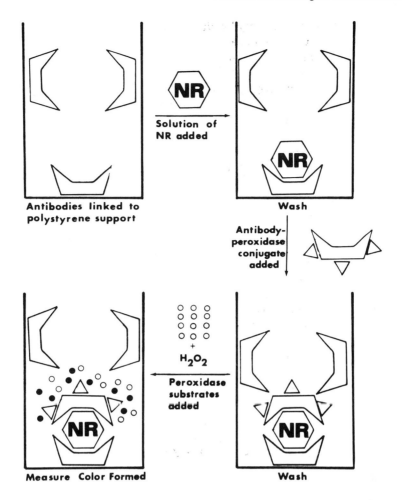

FIGURE 4. Enzyme-linked immunosorbent assay (ELISA) for nitrate reductase. This diagram shows the procedure for doing the ELISA for NR as described in Appendix B.[53,124] The monospecific anti-NR is first attached to a polystyrene support (upper left). Next NR samples are incubated in the coated wells and washed free of unbound proteins (upper right). Next the anti-NR-conjugate is incubated in the wells and binds in proportion to the amount of bound NR (lower right). After the unbound conjugate is washed away, the amount of bound conjugate is determined by assaying for peroxidase activity (lower left). The unreacted dye is represented by (O), while the dye after reaction is shown by (●). The assay is quantified by measuring the absorbance of the reacted dye. A standard curve is prepared with homogeneous squash NADH:NR.

difficult to prepare by other methods of purification.[125] The availability of specific immunoaffinity columns could greatly simplify the preparation of many forms of NR. Toward this end of a wider application of immunochemical methods for the study of NR, a detailed description of the ELISA method for detection and quantification of NR has been attached in Appendix B.[53,124]

II. REGULATION OF NITRATE REDUCTASE

A. Overview and Definitions

Nitrate reductase activity in the leaves of higher plants and all other organisms is adaptive and has been found to dramatically increase in the presence of nitrate. This so-called "induction" of NR activity by nitrate is one of the few higher plant examples of substrate induction, a phenomena that is very common in bacteria. The cellular components and processes involved in the control and regulation of NR activity have been extensively studied, but an integrated and well-accepted explanation has not been elaborated to describe the underlying mechanism(s). This lack of definition in our understanding of NR regulation is largely due to the complexity of the factors controlling NR activity. However, a lack of definition for the properties of NR and of sufficiently sensitive probes for the apo-NR has also contributed to the fuzziness of the current understanding. Since some of these requirements are just at this moment being met, new information now being gained could possibly change current views of NR regulation.

In order to establish a clear starting point for these discussions, we will define "induction" of NR activity in the phenomenological sense without regard to the molecular mechanism, which is to say we will not regard nitrate as an "inducer" per se. Since a "derepression" of NR activity has been shown to occur in many organisms when organic nitrogen sources are depleted in the absence of nitrate, it appears more appropriate to view the increases in NR activity as a relief of repressive influences or derepression. However, for higher plants, especially the crop plants, the level of NR activity appears to be most affected by nitrate availability and, while many factors influence the NR activity, reduced nitrogen does not always lead to a repression of NR.[2] This is in contrast to the simpler plants such as Lemna where NR activity is clearly repressed by ammonia and amino acids.[127]

The regulation of NR activity can be divided into two major aspects: control of the amount of active enzyme and control of the activity of active enzyme. The balance of two opposing processes, *de novo* synthesis of NR and degradation of active NR, controls the level of total enzyme.[2] While the level of active enzyme may be influenced by activators and inhibitors, it has been frequently demonstrated that the synthesis of new NR contributes to the increase in activity during induction by nitrate.[2,6,7,127a] The control of activity will depend on availability of substrates (nitrate and NADH/NADPH), but not on their source. So that in light phases NR will utilize electrons shuttled to it predominantly from the chloroplast, but in the dark would draw from the mitochrondrial shuttles. Rhythms in NR activity observed in some studies can be related to rhythms in light cycles, which influence xylem flow of nitrate and mobilization of stored nitrate, as well as to changes in electron availability.[2,6,7] The kinetic studies of NR indicate that the enzyme has a potentially nonlinear response to changes in cellular redox potential and that efficiency of nitrate reducing activity of NR may be optimal at some substrate concentration less than saturation. Thus, to a certain degree, NR is "self-regulated" as a consequence of its two-site hybrid-random, ping-pong kinetic mechanism which is a very flexible type (see discussion before).

It is important to recognize the small amount of protein that is required to produce the activity of the enzyme. Recently, the squash NADH:NR was purified to a specific activity of 110 units/mg protein.[46] Assuming this specific activity represents intact NR, then in the highly induced squash cotyledon from 0.01 to 0.05% of the extractable protein is NR.[46] Many tissues have lower levels of NR and it can be seen that dramatic changes in NR acticity may represent only very small increases in NR protein. For example, in the studies of Ingle[128] it was found that during nitrate induction of radish cotyledons less than 0.1% of the newly synthesized protein could be NR, even when NR activity had increased 20-fold. It can now be estimated, by assuming the radish cotyledon NR-specific activity is the same as squash, that NR protein increased from 0.0007 to 0.01% of the total protein following induction with ammonium nitrate. These nanogram levels of NR are difficult to detect and to study.

B. Induction of Nitrate Reductase Activity

The influence of the following factors on levels of NR induction and activity have been studied: light, mineral nutrition, plant hormones, N source, water stress, temperature, plant age, antimetabolites, and energy sources (sucrose and organic acids).[2-7,129] Protein and RNA synthesis have also been studied by incorporation of radioactive and density labels into active NR and separation of unlabeled and labeled forms.[2,129a,130] Nitrate uptake and its influence on NR activity are reviewed in a separate section of this volume (see Chapter 4).

Since these studies and others, including such influences as phytochrome, have recently been reviewed,[2,6,7,129] we will focus on the affects of ammonium on NR induction in higher plants. Ammonium appears to repress the induction of NR activity in roots,[18,131] while it enhances the activity in green cotyledons, bud tips, seedlings, and leaves.[2,7,18,131-136] These differences might be due to changes in nitrate uptake in the presence and absence of ammonium, but not all studies have found that nitrate uptake is inhibited in the presence of ammonium.[137] However, the effects of ammonium on NR activity induction may depend on the pH of the induction medium, and the concentration of bicarbonate, ammonium, and nitrate.[2,131] For example, in cotton, the root NR activity was unaffected by ammonium concentrations of 3 mM or less if nitrate was 100 mM, but at 3 mM nitrate the addition of any ammonium decreased the enzyme activity found.[131] Also Oaks[18] and co-workers have shown differential effects of ammonium between the developing root tips and the mature roots. In pea bud tips, ammonium was suggested to be stabilizing NR activity and not directly influencing the amount of NR made.[132] In some cases, ammonium in the absence of nitrate will induce NR activity;[7,133,136] however, it should be noted that at low pH, citrate is an inducer of NR and that aromatic nitro groups and some phytohormones also induce NR in the absence of nitrate.[138] Most plant cell cultures have greater NR activity with ammonium in the medium.[2,6,7]

Several factors contribute to the uniqueness of crop plants with regard to the influence of reduced nitrogen on NR activity. First, nitrate is the most common N source for higher plants and the plants appear to be highly adapted to grow on nitrate.[2] The crop plants have evolved away from the selective use of N sources found with algae, fungi, and simpler plants.[2,127] These adaptive differences between higher plants and "lower" plants are to be expected since the higher plants grow in a much more constant environment than these other organisms. Secondly, the process of photorespiration, which is not unique to higher plant leaves but is very active in this tissue, produces a substantial amount of ammonia in mitochondria and results in a large increase in the ammonia and glutamine pools of the leaf. The rates of ammonia production from photorespiratory glycine to serine conversions are estimated to be ten times the rate of ammonia accumulation from nitrate reduction (see Chapter 7, Volume I for more detailed discussion of photorespiration). The influx of ammonia into the leaf might lead to very little change in existing fluxes and might not be noticed by the tissue. The situation is likely to be different in roots where there is no photorespiration. Thus, the differences in the influence of ammonium on leaves and root of higher plants may be a result of differences in their metabolic patterns.

C. Model Systems: Cell Cultures

The whole plant has proven very difficult to use in studying NR induction since many factors complicate the process. While Oaks[18] and co-workers have used agar-cultured corn roots as a model system, many others have focused studies on plant cells in culture.[2] The plant cell culture offers a system that resembles a nongreen cell in the intact plant because it has a defined life cycle with a lag phase, a division phase, and finally, a senescent phase (see Chapter 3 of this volume).[139] Recently, photoautotrophic cell cultures have been grown and NR activity has been studied in one system, which may then more resemble the life cycle of an intact leaf mesophyll cell and may be a useful model.[140,141] Filner and co-workers[130,142-145] have made extensive studies of NR activity in tobacco cell cultures. In these nonchlorophyllus suspension cultures, NR induction is mediated by a balance between *de novo* synthesis and loss of active enzyme,[130] but ammonium and other sources of reduced N repress NR activity induction by nitrate.[146] For cultures of rose cells, Fletcher and co-workers[147-151] have done extensive studies of the influence of ammonium on the growth in nitrate and the induction of nitrate reductase. It was found that ammonium even in amounts as small as 10% of the total nitrogen stimulated the exponential phase of growth to begin earlier and to have a steeper growth rate.[147,150] Addition of ammonium increased the level of NR activity and shortened the time of growth required before maximum activity was attained.[147-149]

For photoautotrophically grown suspension cultures of *Chenopodium rubrum*, NR activity reaches a maximum during initial growth on ammonium even in the absence of measurable nitrate consumption.[140] These light-driven cells must be grown on high CO_2, which suppresses photorespiration, and consequently these cells may have excess capacity to assimilate ammonium and a need for large ammonium and/or glutamine pools. Ammonium and nitrate appear to contribute equally to the total N content of the *Chenopodium* cells, but growth slows and stationary phase is reached with nitrate still present in the spent medium.[140] In both the rose and *Chenopodium* cells, the ammonium-induced increase in NR activity seems to be associated with a general induction of nitrogen metabolism that leads to an early and rapid accumulation of protein.[140,150,151] Quite interestingly, a recent study of NR activity and growth in *Haworthia* callus showed that both the enzyme and cell division could be induced either by phytohormones or γ-irradiation.[152] Thus, it appears that the effects of ammonium in cell cultures can be attributed to general growth influences such as promotion of the synthesis of phytohormone or other essential cell components (inositol) rather than on N metabolism alone. However, it does appear that changes in N metabolism enzymes occur prior to changes in the cell growth parameters.[140,147]

In Mo-deficient cultures of rose cells, apo-NR appears to be synthesized and can be activated by Mo addition in a process not blocked by RNA and protein synthesis inhibitors.[153,154] Studies by Gamborg and co-workers[155,156] have found stimulatory effects of ammonium and some amino acids on induction of NR activity in cell cultures of several different plants. For soybean, Oaks[157] found that ammonium and glutamine blocked the normal nitrate induction of NR. More recent investigations of tobacco cell cultures have shown that growing cells in the absence of a N source leads to a low level of NR activity, which can be further induced by nitrate.[146] Thus, it has been concluded that in some cell cultures NR is derepressed in the absence of reduced N and that this constitutive NR activity is involved in the regulation of NR induction.[146]

D. Genetic Regulation and Nitrate Reductase Mutants

Obtaining mutant plants deficient in their ability to grow on nitrate has been difficult. Some mutants have impaired ability to use nitrate and therefore are resistant to chlorate, which is often used as a tool to select for NR-deficient mutants (see Section I.H.2).[1,99,158-161] Some of these mutants have alterations in the nitrate uptake process, while others are altered in the activity of NR.[99,158-168] Two types of NR-deficient mutants have been generated.[163,165] Mutants of the NR structural gene are called *nia* mutants and are impaired in the synthesis of apo-NR such that the reductase activity is present but dehydrogenase activity is absent, or there is loss of NR activity but retention of both partial activities. Mutants of the Mo-cofactor are called *cnx* and make an apparently normal apo-NR, but have only dehydrogenase partial activity and lack reductase activity. The *cnx* mutants must also be deficient in enzymes such as xanthine dehydrogenase, which appears to use the same Mo-cofactor as NR.[123a,163,165]

An NR-deficient mutant of *Arabidopsis thaliana* has been described which is apparently deficient in synthesis of effective Mo-cofactor and probably should be classed as a *cnx* mutant.[162,163] For tobacco, mutants have been isolated by several different groups,[160] but the most extensive studies thus far have been reported by Muller and co-workers.[14,86-89,160,161] Several different types of the two major *nia* and *cnx* mutants have been described and active NR can be reconstituted in vitro by mixing extracts of the two classes of mutants.[161,164,165] In addition, somatic hybridization of the mutants leads to an expression of in vivo NR activity and nitrate growth.[166] These mutants are recessive and appear to occur in duplicate.[167] The duplication of the NR structural gene could have contributed to some of the difficulty in obtaining NR-deficient mutants.

Warner, Kleinhofs, and co-workers[123,123a,173,174] have isolated mutant barley plants that have decreased NADH:NR activity, but grow on nitrate. These mutants fall into the two classes described before but are called *nar*-1 for *nia* type mutants and *nar*-2 for the *cnx* type.[123] Most of these mutants contain cross-reacting material as determined by rocket immunoelectrophoresis or protection of active NR against inhibition by anti-NR.[123,123a,175] In addition, it was recently reported that some of these mutants contain NAD(P)H:NR activity, which is not present in the wild type plants.[29]

There is little evidence for the presence of regulatory genes such as have been observed for fungi.[31] These genes are called *nir* and regulate both NR and nitrite reductase apoprotein synthesis.[176] However, this does not rule out the existence of such genes, but just attests to the difficulty in selecting for such mutants in higher plants. In breeding experiments with wheat and barley, some evidence exists for genes that regulate the amount of NR activity and these data are discussed separately in this volume (see Chapter 5).[177] The recent investigations with the mutants of tobacco confirm that NR is constitutive as shown by lack of nitrate induction of NR partial activities in some mutants.[86-89,160] The regulation appears to be via the capability of NR to bind nitrate and to form a "self-inducer" state, which is absent from mutants deficient in an NR or NR component with a nitrate binding site. This type of regulation was found in *Aspergillus* by Cove and Pateman and called autogenous regulation.[178]

The Mo-cofactor has been isolated from wild-type and *cnx* mutants of tobacco and it appears to be similar to the pterin found in other NR (see Section I).[86-89] Most of the *cnx* mutants can be activated by addition of Mo, when the *N. crassa nit*-1 NR assay is used.[179] During growth of cells on nitrate, Mo-cofactor reached its peak level 1 day prior to the peak of NR activity.[89] Mutant cell lines could be divided into three classes based on the levels of Mo-cofactor activity and ability to induce this activity.[89] The differential responses of the Mo-cofactor and the apo-NR led to the suggestion of a linkage between derepression of Mo-cofactor synthesis and NR synthesis. The synthesis of Mo-cofactor appeared to depend on the availability of functional apo-NR.[89,179] The synthesis of Mo-cofactor can then also be linked to the autogenous mechanism for regulation of active NR synthesis, which suggests that the inducing effect of nitrate may be mediated by binding to holo-NR and that this complex promotes not only synthesis of apo-NR but also Mo-cofactor.[89,179]

E. Control of the Activity of Nitrate Reductase
1. Factors Affecting Nitrate Reductase Activity

The activity of NR is a function of a complex set of factors that may vary in importance in different plants.[2,6,7,129a] We will consider four factors: enzyme activity turnover, affectors of NR activity, enzyme location in the cell, and substrate availability. The turnover of NR apoprotein might have been considered separately from the enzyme's activity, and can be once studies of turnover provide data on actual apo-NR

Table 5
SPECIFIC MODIFIERS OF NITRATE REDUCTASE ACTIVITY[a]

Inhibitor source	Mol wt (kdaltons)		Partial activity lost	Effect of		pH Optima	Ref.
	Native	Subunit		Chelator	Heating		
Class I: Proteinase Type Inhibitors							
Maize root	75	66	Dh	Sl. Inh	Inh	Alkaline	180—182, 186
Wheat leaf	38		Red	Inh	Inh	Neutral	183
Class II: Binding Type Inhibitors							
Soybean leaf	31	18	Red	None	Inh		117, 191
Rice cells	150	73	Dh	Inh	Inh	Alkaline	180, 184—187
Rice plants	46[b]		Dh		Inh		186, 188—190
Rice roots			Red		Inh		192, 193

[a]   Abbreviations: Dh, dehydrogenase partial activity; Red, reductase partial activity; Sl. Inh, slight inhibition; Inh, total inhibition.

[b]   If the inhibitor were bound to native NR (M_r = 210 kdaltons), its M_r was 240 kdaltons.

protein and not just on NR activity. In any case, it is difficult to divorce the turnover of NR from the expressed activity of the enzyme.

2. Turnover of Nitrate Reductase Activity

The level of NR activity in plant tissues and cell cultures appears to result from simultaneous synthesis and degradation of active NR.[2,6,7,129a,130] The changing balance of these opposing processes generates the NR activity increases initially found during induction, but it is not clear what sort of changes occur in the balance during arrival at the steady-state level of NR activity. The evaluation of the steady-state rate of turnover has been most often done by blocking synthesis of active NR and measuring decay of NR activity.[2,6,7] Loss of NR activity in higher plant tissues can be initiated in several ways. Tungstate has been studied as an antagonist of molybdate and these data have been recently reviewed.[1] The effect of tungstate is on newly synthesized NR and, thus, provides for means to study the decay of existing holo-NR.[1] Inhibition of RNA and protein synthesis may permit study of holo-NR decay, but these inhibitors may also prevent the synthesis of inactivation systems directed at the active NR.[2,6,7] Placing plants in continuous darkness leads to a decline in NR activity and N-starvation can also trigger NR loss. Inactivation of NR can involve either covalent and irreversible changes of structure such as occur upon protease degradation of the enzyme, or noncovalent binding of modifiers.[1,2]

Two lines of data indicate that holo-NR is proteolytically degraded: recent isolations of "nicked" NR subunits in highly purified barley and squash NADH:NR;[15,46] and the isolation of proteases with high specificity for inactivating NR.[2,6,7] Wallace, Oaks, and co-workers[18,75,180,181] have extensively studied a protein isolated from the roots of corn. The properties of this factor are presented in Table 5 along with those of other protein factors that inhibit NR. The corn root inhibitor has all the characteristics of a proteolytic enzyme and has been shown to degrade *C. vulgaris* NR into fragments.[180-182] A rather nice comparison of this factor to the rice cell inhibitor showed the irreversible action of corn factor on corn and rice NR.[180] The NR inactivator from wheat has some properties like a protease but did not yield small fragments when inactivating highly purified wheat NADH:NR.[183] Proteolytic degradation of the holo-NR to peptide fragments and domains is clearly a mechanism used in some plant tissues to inactivate NR, but the regulation of these irreversible processes is only vaguely understood.[15] In actuality, the process of NR proteolysis has not yet been systematically studied and it is not known if introduction of "nicks" leads to alteration of NR activity. In addition, it may be important to know the progression of NR activity change with degree of polypeptide degradation, since current studies indicate high activity for partially degraded NR.[15,46,48]

3. Nitrate Reductase Affectors

Cyanide has been implicated as a regulator of holo-NR in *C. vulgaris* and some other algae, where cyanide binds tightly to the nitrate site of reduced enzyme and activation occurs by oxidation of the reduced-enzyme cyanide complex.[194] This mechanism has not been shown for higher plant NR, although some plant NR may be activated by oxidative treatments.[1] Most often activation of plant NADH:NR has been found with cysteine, other thiols, or EDTA.[1,28a,50,52,195] We have studied these activators and concluded that "activation" is a chelator effect and functions via the chelation of heavy metals in assay mixtures.[50,52,53] We have found that in order to study NR in the absence of chelators, heavy metals must be removed from the buffers.[46] It appears that copper leads to oxidation of NR thiols and that cysteine can reactivate this copper-inhibited NR when EDTA cannot.[50,53] The inhibition of NR by zinc appears to be reversible and inhibition may only result from binding of the heavy metal.[46,50,53] It does not seem reasonable to attribute physiological significance to metal ion inhibition at our current level of understanding.

Yamaya and Ohira,[184-187] and Yamaya et al.[180] have extensively studied a high molecular weight NR inhibitor that reversibly inhibits NADH:NR and is found in the cell cultures of rice and many other plants (Table 5). The highest levels of inhibitor could be correlated with low levels of NR activity in crude extracts of rice cells, and the purified inhibitor binds to NR in a process that can be overcome by NADH.[184,187] This factor appears to be inhibited by mercurials indicating it may depend on thiol functional groups.[185] Jolly and Tolbert[191] isolated from the leaves of soybean, a lower molecular weight NR inhibitor that was light sensitive and present in larger amounts in the dark.[117,191] This factor was a noncompetitive inhibitor of nitrate for the soybean NADH:NR but had no effect on the NAD(P)H:NR. The soybean factor was not dependent on thiol functional groups, which might be correlated with the need for thiol reducing agents during the isolation of soybean leaf NADH:NR (soybean NAD(P)H:NR can be isolated in the absence of thiol).[9,191] It has also been suggested that rice roots contain a protein inhibitor of NR, which may be the peroxidase of that tissue.[192,193] These factors are not inactivating NR by proteolytic degradation but all appear to be inhibiting by binding to NR in a reversible manner (Table 5).[180]

Also Leong and Shen[188-190] have studied a binding inhibitor specific for NR, which is found in the roots and other tissues of rice seedlings. They found that the inhibitor present in root extracts is present in both free and bound forms, which had a low and a high M_r, respectively.[190] These two forms of rice root inhibitor were separated by affinity chromatography on blue Sepharose, which showed that the high M_r form was a combination of the low M_r inhibitor and native NR.[190] In kinetic studies, the free inhibitor blocked NR activity in two phases, where initial binding led to partial inhibition of both NADH:NR and methyl viologen reductase activities. However, after a 20-min incubation, all the NADH:NR activity is lost without further decrease in the reductase activity.[190] Leong and Shen[188,190] concluded that the rice root inhibitors action was against the dehydrogenase function of the NR, but also found that the $FADH_2$-linked reductase activity was inhibited in parallel with the cyt. c reductase activity (which was the assay they used for measuring the dehydrogenase activity).

Most interestingly, rice root inhibitor in the form bound to NR retained some ability to inactivate additional freshly added NR, indicating the possibility of a polyvalent character for this NR inhibitor.[190]

While the two types of protein inhibitors, namely the NR proteinases and the NR binding-proteins, appear in some cases to be physiologically active, these factors may be sequestered in the cell and separated from NR.[2,6,7] In fact, corn leaves contain no proteinase activity similar to the activity found in the corn roots.[180,181] Thus, the presence of an NR inhibitor in plant tissue does not establish it as an agent for regulation of enzyme activity. This must also be said for the protein activators of wheat leaves and the ethanol-extractable activator of corn scutella.[196,197] So while affectors that bind to NR and reversibly inhibit activity cannot be assigned a clear role in regulation of enzyme activity, the NR proteinase factors are a real problem concerning investigators interested in the study of high purity NR. These proteinase factors appear to exert influence even in plants with low endogenous decay of NR like squash and spinach, which yield nicked holo-NR.[1,15,48,74] The isolation of partially degraded subunits of NR has obscured the description of apo-NR.[1,15,48] Further study of NR proteinase is needed, which would include considerations of the influences of these agents in vivo and interactions with nitrogen nutrition. Even if these agents are not active in vivo, the effects they exert during isolation of NR make it important for effective inhibitors to be found, which might greatly aid the study of NR in higher plants.

4. Subcellular Location of Nitrate Reductase

Differential centrifugation of cell extracts and organelle isolations tends to indicate that NR is a soluble enzyme in the cytoplasm of both leaf and root cells.[1,2,33] Some studies of these two tissues have disagreed and shown NR bound into organelles or latent in some membrane fractions.[33] Two general observations are that NR is very easy to extract from most tissues, despite the complex requirements for protectants,[33,50,52,53] and that purified enzyme is also sticky and will bind to hydrophobic affinity columns.[26,52] Several investigators have suggested that NR may be associated with membranes in the cell;[1,2] the wide variety of artificial activities catalyzed by NADH:NR tends to indicate that the enzyme might be membrane bound or otherwise complexed in the cell.[41,195] The attachment of NR to chloroplast envelopes or other membranes might be rather loose or easily broken.[1,2,47] Recently, two different histochemical studies and one immunochemical localization of NADPH:NR in *N. crassa* have indicated a membrane-bound or vesicular site for NR.[198-200] Similar studies have not been done with higher plant leaf NR as yet. If NR is membrane bound or sequestered in some manner in the cell, this would be an important consideration for enzyme induction and regulation.

5. Substrate Availability
a. Nitrate

The uptake of nitrate and the regulation of that process is being discussed elsewhere in this book (see Chapter 4). This substrate has a complex character because it is reduced by NR, but also affects the level of enzyme activity by "inducing" it.[2] By movement in the transpiration stream nitrate enters the leaf in light, but also may be mobilized from the now established location in the vacuole of the leaf cell.[129a,201,202] The metabolically active nitrate pool then may be fed from either the xylem stream or that stored in the vacuole. Many studies have found the plant incapable of total reduction of all nitrate present, and tissues senesce and die while still containing nitrate.[2] Interestingly, the cell cultures of higher plants sometimes senesce in the presence of nitrate, which may be especially significant in the *C. rubrum* photoautotrophic cell culture where the energy source was not limiting.[140,147] If nitrate is the only oxidant in vivo that can react with reduced NR, the availability of nitrate would control catalytic turnover of the enzyme and might influence stability since over-reduced enzyme is labile.[1,2,47] However, iron-chelates may oxidize NR and maintain a certain degree of catalytic turnover among NR molecules even in the absence of nitrate.[74] Some interesting studies have been done of the interaction of nitrate and iron metabolism.[105,203]

b. NADH, NADPH, and Other Electron Sources

Much of the study of electron donors for NR has focused on NADH availability since the NADH:NR form is most common in higher plants.[2,6,7] However, a recent review discussed electron sources for nitrate reduction in nongreen tissues, which contain both NADH:NR and NAD(P)H:NR.[204] A large number of investigations have indicated that nitrate reduction is most active in the light and halts in darkness.[205] The metabolic mechanism suggested to account for this is the competition for electrons between mitochondrial electron transport and NR, which shifts in light to dark transition because the high ATP level in light inhibits mitochondrial oxidations.[6,7,205] The recent results of Stitt et al.[206] show that the cytoplasmic ratio of ATP:ADP is about the same in both light and dark and that high energy charge may not be a regulator of mitochondrial activity. In addition, recent studies have shown that nitrate is reduced in the leaves of barley in the dark, but this dark reduction is slower than in light.[23,129a,207]

NADH may be provided to the NR via several different reactions in the cell and these sources may differ between the light and dark phases of growth.[6,7,23,129a,207] The most logical way to view this is to consider what substrates and enzymes are involved in feeding electrons for the NADH pool in general, since NR will not be able to distinguish among NADH molecules. Clearly, in the light the ultimate source of electrons are the light reactions in the chloroplasts and it should be understood that the chloroplast is probably the largest contributor of electrons for nitrate reduction in the light.[3] This may be carried out via the triose-phosphate shuttle or the malate/asparate shuttle.[2,3] Production of malate in the mitochondria in the light might also contribute via malic enzyme or malate dehydrogenase, and this might be a distinct source of electrons in consideration of the large store of citrate found in vacuoles.[6,7] Malate may also come from the vacuoles and be directly used in the cytoplasm for the production of reducing power. In the dark, the pattern shifts and chloroplast and all other leaf-reducing systems are energized by the collective action of the mitochondria.[23,129a,207] Thus, the same reactions and substrates may be important in the light and dark, but the ultimate source of electrons shifts from the chloroplast in the light to the mitochondria in the dark. In roots and other nongreen organs, the mitochrondria are the source of substrates for energy with the phloem providing many of the substrates to be used either directly by extramitochondrial systems or indirectly via mitochondrial activities.[204]

NADH has been suggested to be involved in the regulation of NADH:NR in two ways.[1-7] NADH may act as an activator of NR inactivated by a binding-protein inhibitor, such as has been described from rice cell cultures.[180] Alternatively, NADH may act as an inactivator of NR by reducing it in the absence of electron acceptors, which could create an internal instability (i.e., over-reduction of the enzyme's redox centers) and loss of enzyme activity.[1,2,47] The inactivation of NR by NADH is markedly enhanced in the presence of heavy metal ions.[50,52,53] The balance among these effects of NADH on NR is unknown; however, the corn root NR proteinase has been shown to effect the oxidized and NADH-reduced NR of corn leaf and rice cells to the same extent.[180] Instability of NR caused by reduction in the absence of electron acceptors (i.e., nitrate) remains to be further investigated.

F. An Integrated Model of Nitrate Reductase Regulation

The fragmentary information available on NR regulation does not permit integration into a single coherent model. We have already identified some of the gaps in the knowledge needed for this model such as data on apo-NR synthesis, holo-NR degradation, and subcellular location of NR. It appears that regulation differs for green and nongreen tissues and may vary among the various plant types.[2,3,129a] The presence of NAD(P)H:NR forms in some tissues complicates regulation and if other forms of NR can be expressed in some tissues, these complications may be worse than is currently perceived. This type of complicated development pattern can be seen in the differential induction found for NAD(P)H:NR and NADH:NR in the soybean cotyledon.[44] Many studies have focused on the effects of Mo and W on NR activity,[1] but it has also been shown that appearance of NR activity can be blocked by inhibiting heme synthesis.[208] The uniqueness of the Mo-cofactor to NR and a few related Mo-containing enzymes offers opportunities for study of coordinated synthesis of both the apo-NR and its cofactors,[89] which appears to be an area where immunochemical methods such as the ELISA could be used to detect NR proteins and their domains.[53,124] Immunochemical analyses have already begun to contribute to the study of NR mutants,[123,123a,170] but could also prove useful here to isolate incomplete apo-NR by immunoaffinity chromatography. It may be that the understanding of NR regulation will depend on a more general understanding of the regulation of metabolism in higher plant leaves and roots, and better understanding of higher plant genetic mechanisms. The specific proteinases found in yeasts have been suggested as important factors in regulation of metabolism in these organisms.[209] Regulation of proteinase activity in higher plants is under study also.[75,129a,181,183] Here again, immunochemistry might offer powerful tools for investigating amounts of these degradative factors and could be useful in determining cellular location. However, plants selected for high levels of stable NR activity could prove useful in many aspects of study.

G. Molecular Biology of Nitrate Reductase

The genetical and molecular biological control of nitrogen metabolism in higher plants is poorly understood. Extensive study, however, has been reported for the genetic regulation of nitrogen metabolism in the fungi *N. crassa* and *Aspergillus nidulans,* and these studies have been reviewed elsewhere.[210] The majority of the knowledge has been derived from the study of mutants which specifically alter the pathway of nitrogen acquisition. In *N. crassa,* the *nit*-3 gene encodes for the NR apoprotein. Several genes are involved in Mo-cofactor production (*nit*-1,7,8,9). In addition, *nit*-4 mediates nitrate induction, while *nit*-2 is involved in metabolite repression. Further, it has been demonstrated that NR autogenously regulates the nitrate assimilatory pathway in both *N. crassa* and *A. nidulans.*[176,211]

Based on the above fungal models, a genetic approach to understanding nitrogen metabolism in higher plants has been reported using tobacco cell lines, *Pisum sativum* and *Arabidopsis thaliana,* baced on the generation of mutants unable to use nitrate as a nitrogen source and resistance to the nitrate analogue chlorate.[158-168,212] In addition, NR mutants in barley have been screened by monitoring enzyme activity by an in vivo NR assay.[173-175] Muller and co-workers[87-89,163] have described two distinct classes of mutant cell lines in tobacco. The *nia* mutants are apparently analogous to the *N. crassa nit*-3 mutant, which both encode for defective NR apoproteins.[163] The NR activity of the second type *(cnx)* can be restored by reconstitution with various Mo-cofactor preparations, indicating that the *cnx* class of mutants is defective in active Mo-cofactor like *N. crassa nit*-1.[87-89] Kleinhofs, Warner, and co-workers[123,123a,175] have shown that barley mutants contain cross-reacting material using anti-barley NR, and that one gene encoded for the NR apo-protein (*nar*-1) and another gene encoded for the Mo-cofactor (*nar*-2). The *nar*-1 mutants were able to grow on nitrate as sole nitrogen source, and some contain an NAD(P)H:NR activity.[29,174] It was concluded that the mutants contained a second bispecific form of NR, but that this form was immunologically not closely related to wild-type NR.[29,123,123a] However, no evidence was found for this NAD(P)H:NR in the extracts of the wild-type.[113] Regulatory mutant such as *N. crassa nit*-2 and *nit*-4, have been found among the higher plant NR-deficient mutants.[211]

The molecular biology of nitrate assimilation is an area of a great deal of current research. If the subunit M_r of higher plant NR is 110 kdaltons, then an approximately 3-kilobase fragment of DNA would be required to encode for this polypeptide chain (disregarding introns and upstream promoter regions). Plant genetic manipulation has the potential to increase the levels of NR, generally considered the rate-limiting enzyme of nitrate assimilation. It has been suggested that if the gene from a plant with an efficient NR could be substituted for one from a plant with a less efficient NR, increased yield could be realized.[213] Somatic hybridization has already resulted in the fusing of protoplast of NR-deficient tobacco cell lines, which yielded active NR after extraction.[171] Although efficient higher plant transformation systems have not yet been reported, it can be expected that one will soon be forthcoming. Gene cloning appears to be the means both to understand better higher plant nitrate metabolism and perhaps to increase the efficiency of this process. Delineation of the fungal gene for NR may provide a means to extract the higher plant NR gene, since the NR proteins from fungi and higher plants have been shown to share a degree of sequence homology when analyzed by immunological methods.[32] Another approach to the isolation of the NR gene would be to use anti-NR to precipitate the nascent polypeptide chain for NR during its biosynthesis while still attached to the ribosome. If the highly unstable m-RNA for NR could be isolated, c-DNA could be cloned and the polynucleotide sequenced. Alternatively, if a partial amino acid sequence for higher plant or other NR were determined, these data could be used to generate "fish-hook" nucleotide probes for isolating the NR gene. To date, no reports have been made concerning the isolation of the higher plant NR gene.

III. APPENDIX

A. Affinity Chromatography of Nitrate Reductase

1. Blue Dextran Sepharose

Blue dextran is coupled to cyanogen bromide-activated Sepharose 4B as described by Ryan and Vestling,[214] with the modifications described by Sherrard and Dalling.[12] This affinity gel was first used by Solomonson[215] for the purification of *C. vulgaris* NADH:NR where he eluted the enzyme with a gradient of NADH. Subsequently, NR forms from spinach, wheat and barley leaf extracts have been purified with blue dextran as the affinity ligand.[10,12,13,216] Campbell[9,28] and Orihuel-Iranzo and Campbell[44] have used blue dextran Sepharose to separate and partially purify the NADH: and NAD(P)H:NR forms found in the extracts of soybean leaves and cotyledons, and corn scutella. However, in general, application of higher plant NR to blue dextran Sepharose results in a large percentage of apparent nonspecific binding, and only partially achieves the success obtained with *C. vulgaris* NR.[52,217]

2. Blue Sepharose

Cibacron® blue F3G-A (Polysciences Inc., Warrington, Pa.) can be coupled directly to Sepharose CL-4B without cyanogen bromide activation.[11,218] Deionized water-washed Sepharose is suspended in 350 mℓ of water and heated to 60°C. A 60-mℓ solution containing 2 g Cibacron® blue is added dropwise, and stirring continued for 30 min. Forty-five (45) g NaCl is added and stirring continued for 1 hr. The mixture is subsequently heated to 80°C , and after 4 g Na_2CO_3 is added, the stirring is continued for 2 hr more. After cooling to room temperature, the gel is washed on a Buchner funnel with water until the filtrate becomes colorless. The gel is stored at 4°C until needed. Blue Sepharose is also commercially available from several sources. Initially, NADH:NR forms from corn leaves and squash cotyledons were purified using a 0.1-mM NADH elution from blue Sepharose.[11] Barley NADH:NR was also purified using blue A Sepharose (Amicon).[56] The NADH: and NAD(P)H:NR forms found in the extracts of roots were separated and purified with blue Sepharose.[25] Mendel[219] has made a comparison of affinity gels for the purification of tobacco NADH:NR and NADH:cyt. c reductase activity found in mutant tobacco cell lines. He found that blue dextran Sepharose was superior to blue Sepharose or AMP Sepharose, but could only achieve specific activity of 1 unit/mg protein.[219] Although providing greater than 100-fold purification, the higher plant NR preparations made by specific elution from blue Sepharose appear to be only 10 to 20% pure when evaluated by polyacrylamide gel electrophoresis.[41]

3. Pink Sepharose

Procion red HE3B is coupled to Sepharose CL-4B by a similar procedure as used for blue Sepharose, but a lower temperature is used such that less dye is coupled to the gel.[220] Pink Sepharose was prepared after red Sepharose, which was made in a procedure identical to the blue Sepharose synthesis except that Procion red was substituted for Cibacron® blue, was found to bind NR too tightly for efficient elution with NADH.[47,53] Mendel[219] found a similar result for blue Sepharose when purifying tobacco NADH:NR and achieved a satisfactory gel by diluting blue Sepharose with unactivated gel in a 1:5 ratio. After concentration of blue Sepharose-purified NR by ammonium sulfate precipitation, the squash enzyme is bound to pink Sepharose, unbound protein washed away with buffer, and the highly purified NR can be eluted with 0.1 mM NADH in buffer.[47,48,53] Additional NR can be eluted by treating the column with 0.1 mM FAD, in some cases, and with 0.3 M KNO_3.[48] The squash NADH:NR purified using blue Sepharose, gel filtration on Bio-gel® A 1.5m, and pink Sepharose (NADH eluted only) had a specific activity of 40 to 50 units/mg protein.[48] However, while this enzyme is homogenous, it appears to be largely composed of cleaved subunits, which were found to have a M_r of 75 and 45 kdaltons.[48]

4. Zinc Chelate Affinity Chromatography

A zinc chelate affinity matrix can be prepared as described by Porath and co-workers.[221] This affinity media consists of essentially half EDTA molecules attached to Sepharose, to which Zn^{2+} has been chelated. Synthesis involves the coupling of iminodiacetic acid to oxirane-activated Sepharose, with subsequent loading of the zinc.[50,221] Immobilized iminodiacetic acid can be obtained from Pierce Chemical Company or Pharmacia Fine Chemicals. Squash cotyledon NR was shown to be bound to and eluted from a Zn^{2+} column by EDTA.[40,53] However, high specific activity homogeneous squash NADH:NR (110 units/mg protein) could be obtained by application of blue Sepharose purified to a zinc chelate column and elution with a lowered pH.[46] Squash cotyledons are extracted without adding cysteine as has been described,[50] and purified on blue Sepharose by specific elution as before.[11] The NR bound to blue Sepharose is washed with a phosphate buffer containing 10 μM FAD and EDTA is omitted from the final three washes out of a total of seven washes.[46] In order to elute NR from the blue Sepharose column with NADH in the absence of EDTA, the buffers to be used must be metal-free. Metal-free buffers were prepared by passing them over 8-hydroxyquinoline-controlled pore glass bead columns (Pierce Chemical Company).[46] High activity fractions from the Blue Sepharose column are directly applied to 2 mℓ

Zn^{2+} column, which has a capacity for binding NR of about 100 units/mℓ of gel.[46] The NR bound to the Zn^{2+} column was washed with 75 mM K-PO$_4$, pH 7, containing 1 M NaCl and 10 μM FAD. The NR activity was eluted with the same buffer, pH 6.2, with a recovery of 40 to 50%. This NR was 3300-fold purified and resulted in a yield of 0.5 mg/kg 5-day-old squash cotyledons. The procedure requires about 10 hr to complete.[46] These preparations contain both intact 115-kdalton subunit apo-NR and some lesser amounts of cleaved subunits.[74]

B. Enzyme-Linked Immunosorbent Assay (ELISA) for Nitrate Reductase[53,124]

1. Materials

Polystyrene 96-well microtiter plant with flat-bottomed wells — Microtiter plates certified to be free of edge-effects (spurious readings near edges of plate possibly due to residual plasticizer) and with background readings of +/- 0.005 absorbance units from the mean can be obtained from Nunc (Nunc Immuno-Plate I).

Phosphate-buffered saline (PBS) and phosphate-buffered albumin (PBA) — 40 mM K-PO$_4$, 0.15 M NaCl, pH 7.5 was used as PBS, to which 0.1% (v/v) Tween-20 was added to prepare PBS-Tween. To prepare PBA, bovine serum albumin is added to PBS at 1.0% (w/v). It should be established that the albumin to be used is free of peroxidase activity.

Purified monospecific antibody against nitrate reductase — The IgG fraction of the monospecific anti-NR should be separated from the whole serum as described below. If unfractionated anti-NR is to be used, it must be purified by ammonium sulfate precipitation and then dialyzed against PBS.

Peroxidase — Horseradish peroxidase Type VI was purchased from Sigma Chemical Company, St. Louis, Mo. The peroxidase to be used should have an activity RZ value of at least 3.

2,2′ azino-di(3-ethyl-benzthiazoline-6-sulfonate) disodium salt (ABTS) — The substrate solution contained 0.1 M citrate, pH 4.0, 2 mM ABTS and 4 mM H$_2$O$_2$. Stock solutions of ABTS must be stored in the dark and refrigerated. The hydrogen peroxide should be from a fresh bottle of a 30% solution.

2. Preparations

Purification of IgG — The IgG fraction of the whole anti-NR serum may be effectively separated from other serum fractions by collecting the void volume fraction that passes through a DEAE Affi-Gel® Blue column (Bio-Rad Laboratories).[124,222] This procedure frees the IgG fraction of all contaminants except transferrin and can also remove most of the proteolytic activity found in serum. Less effective purification can be done by combining passage of the serum successively over blue Sepharose and then DEAE-cellulose or -Sephadex.

Antibody coating of the microtiter plates — The anti-NR is diluted with PBS to a concentration of 10 μg/mℓ if the IgG fraction is used or to 0.1 to 0.2 mg/mℓ if ammonium sulfate-fractioned serum is used.[53,124] Of this diluted anti-NR solution, 0.2 mℓ is added to each well and incubated for 3 hr at room temperature. These plates are then stored in the refrigerator overnight. Longer times of storage will depend on the stability of the anti-NR and must be determined in each case.

Conjugation of IgG and peroxidase — The purified IgG fraction of the monospecific anti-NR was coupled to horseradish peroxidase by the periodate method of Wilson and Nakane.[223] This procedure can only be used if serum has been purified to obtain the IgG fraction.[124] If only ammonium sulfate-fractionated serum is to be used, then the coupling procedure described in the following paragraph should be used. A 1-mℓ solution containing 4 mg of peroxidase in distilled water was mixed with 0.2 mℓ of freshly prepared 0.1 M NaIO$_4$ for 20 min at room temperature. Then the solution was dialyzed overnight against 1 mM sodium acetate buffer, pH 4.4, in a cold room. The pH of the peroxidase-aldehyde solultion was adjusted to 9 to 9.5 by adding 0.02 mℓ of 0.2 M sodium carbonate, pH 9.5, and immediately mixed with 4 to 6 mg of IgG in 1 mℓ of 0.01 M sodium carbonate, pH 9.5. It should be established that the IgG fraction is free of ammonium by dialysis vs. 0.01 M carbonate buffer prior to the coupling step. The coupling mixture is stirred for 2 hr at room temperature. From a freshly prepared solution of sodium borohydride, 4 mg/mℓ deionized water, 0.1 mℓ was added and the solution was left for 2 hr in a refrigerator. The conjugated mixture was gel filtered on Sephacryl® S-200 (Pharmacia Fine Chemicals) using 10 mM NaPO$_4$, 0.15 M NaCl, pH 7.0 for equilibration. The absorbance of the 2-mℓ fractions was read at 280 and 403 nm and those fractions having an absorbance ratio for A$_{403\ nm}$:A$_{280\ nm}$ of 0.3 to 0.6 were pooled. Ten (10) mg/mℓ of albumin is added to the pooled fractions of "conjugate", which were stored in the freezer.

Conjugation of ammonium sulfate-purified anti-NR to peroxidase — The two-step glutaraldehyde method of Avrameas and co-workers[224] was used. Ten (10) mg of peroxidase was dissolved in 0.2 mℓ of 1% glutaraldehyde solution in 0.1 M K-PO$_4$, pH 6.8 and allowed to react for 18 hr at room temperature.[53] Unreacted glutaraldehyde is removed by gel filtration on Sephadex® G-25 equilibrated with 0.15 M NaCl. The brown peroxidase fractions were pooled and mixed with 1 mℓ of anti-NR ammonium sulfate-purified IgG (5 mg/mℓ in 0.15 M NaCl). The pH was adjusted by the addition of 0.2 mℓ of 0.5 M sodium carbonate, pH 9.5, and the solution left for 24 hr in the refrigerator. To the peroxidase/IgG solution, 0.1 mℓ of 1 M lysine, pH 7, was added and the solution left for 2 hr more in the refrigerator. The conjugate was purified by ammonium sulfate precipitation and stored in PBS in a freezer.

3. Assay[53,124]

1. The coated microtiter plate is allowed to come to room temperature and the wells are aspirated.
2. After washing the plate three times with PBS-Tween, each wall is incubated with 0.2 mℓ PBA-Tween for 50 min at room temperature.
3. After the wells are aspirated, 0.1 mℓ NR samples are incubated for 2 hr in a humid atmosphere at 37°C. The samples of NR must be diluted to the range of 0.1 to 0.01 units/mℓ (units = μmol/min) with 0.1 M KPO$_4$, 1 mM EDTA, pH 7.5, or the buffer used to extract the tissue (best to omit thiol reagents from this dilution buffer). For controls at least six wells should be simultaneously incubated with only PBA or the dilution buffer.
4. The NR samples are aspirated from the wells and each well is washed three times with PBA-Tween allowing the wash to remain in the well for 2 to 5 min.
5. An aliquot of 0.1 mℓ of conjugate, which was appropriately diluted in PBS-Tween, was incubated in the wells for 2 hr at room temperature. For the conjugate prepared with pure IgG (see page 38, "Conjugation of IgG and peroxidase"), a dilution of 1:200 can be used. For partially purified IgG conjugates (see "Conjugation of ammonium sulfate-purified anti-NR to peroxidase"), a dilution of 1:50 may be required. Optimum dilution of the conjugate can be determined by allowing purified NR to bind to a microtiter plate overnight and then incubating serially diluted conjugate in the wells and determining the least amount of conjugate to give a reliable color response in the peroxidase assay.
6. Wash the wells three times with PBS-Tween, using 2- to 5-min incubations for each wash.
7. Incubate 0.12 mℓ of substrate solution in each well for 50 min at room temperature in the dark.

8. From each well, 0.1 ml is quantitatively removed and diluted to 1.0 ml with 0.1 *M* citrate, pH 2.8. The absorbance of these solutions is read immediately at 410 nm. The pH of the citrate used for dilution should not be allowed to be below 2.8, because more acidic solutions appear to have high rates of increase in absorbance upon standing.

9. After correction for background, a standard curve can be made using a selected NR standard and the amounts of NR present in unknowns can be determined. Please note: it may not be surprising if the ELISA color response is stronger per unit of NR activity in crude extracts than purified fractions of NR.[53,124] This is due to the presence of fragments of NR which are present in the crude extract and separated by blue Sepharose.[124] These fragments of NR can be found in the wash of proteins which do not bind to the affinity gel.

ACKNOWLEDGMENTS

We would like to thank the National Science Foundation, which has supported the efforts of our laboratories via grants to W.H.C. (PCM 76-18803, PCM 79-15298, PCM 83-02146, and DMB 85-02672) and to J.S. (DMB 84-04243), and the U.S. Department of Agriculture, Competitive Grants Office, which has provided funds to W.H.C. via grants 83-CRCR-11289 and 85-CRCR-11681. We would also like to thank S. V. Evola, C. Neyra, M. G. Redinbaugh, and K. Ripp for permission to use their unpublished results. Finally, we thank M. G. Redinbaugh, K. Ripp, W. B. Mahony, B. Orihuel-Iranzo, C. Neyra, and R. H. Garrett for discussions pertaining to this review and the ideas expressed herein.

REFERENCES

1. Hewitt, E. J. and Notton, B. A., Nitrate reductase systems in eukaryotic and prokaryotic organisms, in *Molybdenum and Molybdenum-Containing Enzymes,* Coughlin, M. P., Ed., Pergamon Press, Oxford, 1980, 273.
2. Beevers, L. and Hageman, R. H., Nitrate and nitrite reduction, in *The Biochemistry of Plants,* Vol. 5, Amino Acids and Derivatives, Miflin, B. J., Ed., Academic Press, New York, 1980, 115.
3. Guerrero, M. G., Vega, J. M., and Losada, M., The assimilatory nitrate-reducing system and its regulation, *Annu. Rev. Plant Physiol.,* 32, 169, 1981.
4. Vennesland, B. and Guerrero, M. G., Reduction of nitrate and nitrite, in *Encyclopedia of Plant Physiology,* New Series, Vol. 6, Photosynthesis II, Photosynthetic Carbon Metabolism and Related Processes, Gibbs, M. and Latzko, E., Eds., Springer-Verlag, Berlin, 1979, 425.
5. Losada, M., Guerrero, M. G., and Vega, J. M., The assimilatory reduction of nitrate, in *Biology of Inorganic Nitrogen and Sulfur,* Bothe, H. and Trebst, A., Eds., Springer-Verlag, Berlin, 1981, 30.
6. Srivastava, H. S., Regulation of nitrate reductase activity in higher plants, *Phytochemistry,* 19, 725, 1980.
7. Naik, M. S., Arrol, Y. P., Nair, T. V. R., and Ramarao, C. S., Nitrate assimilation — its regulation and relationship to reduced nitrogen in higher plants, *Phytochemistry,* 21, 495, 1982.
8. Evans, H. J. and Nason, A., Pyridine nucleotide-nitrate reductase from extracts of higher plants, *Plant Physiol.,* 28, 233, 1953.
9. Campbell, W. H., Separation of soybean leaf nitrate reductases by affinity chromatography, *Plant Sci. Lett.,* 7, 239, 1976.
10. Notton, B. A., Fido, R. J., and Hewitt, E. J., The presence of functional haem in a higher plant nitrate reductase, *Plant Sci. Lett.,* 8, 165, 1977.
11. Campbell, W. H. and Smarrelli, J., Jr., Purification and kinetics of higher plant NADH:nitrate reductase, *Plant Physiol.,* 61, 611, 1978.
12. Sherrard, J. H. and Dalling, M. J., In vitro stability of nitrate reductase from wheat leaves, *Plant Physiol.,* 63, 346, 1979.

13. Kuo, T., Kleinhofs, A., and Warner, R. L., Purification and partial characterization of nitrate reductase from barley leaves, *Plant Sci. Lett.,* 17, 371, 1980.
14. Mendel, R. R. and Muller, A. J., Comparative characterization of nitrate reductase from wild-type and molybdenum cofactor-defective cell cultures of *Nicotiana tabacum, Plant Sci. Lett.,* 18, 277, 1980.
15. Wray, J. L., Structure of higher plant nitrate reductase, presented at Int. Symp. Nitrate Assimilation — Molecular and Genetic Aspects, Gatersleben, G.D.R., June 22 to 24, 1982, 6.
16. Aslam, M., Differential effect of tungsten on the development of endogenous and nitrate-induced nitrate reductase activities in soybean leaves, *Plant Physiol.,* 70, 35, 1982.
17. Ryan, S. A., Nelson, R. S., and Harper, J. E., Nitrogen assimilation and inheritance of three soybean mutants with decreased leaf nitrate reductase activity, *Plant Physiol.,* 69S, 116, 1982.
17a. Timpo, E. E. and Neyra, C. A., Expression of nitrate and nitrite reductase activities under various forms of nitrogen nutrition in *Phaseolus vulgaris* L., *Plant Physiol.,* 72, 71, 1983.
18. Oaks, A., Nitrate reductase in roots and its regulation, in *Nitrogen Assimilation of Plants,* Hewitt, E. J. and Cutting, C. V., Eds., Academic Press, London, 1979, 217.
19. Jackson, W. A., Nitrate acquisition and assimilation by higher plants: processes in the root system, in *Nitrogen in the Environment,* Vol. 2, Soil-plant-nitrogen relationships, Nielsen, D. R. and Mac-Donald, J. G., Eds., Academic Press, New York, 1978, 45.
20. Pate, J. S., Uptake, assimilation and transport of nitrogen compounds by plants, *Soil Biol. Biochem.,* 5, 109, 1973.
21. Crafts-Brandner, S. J. and Harper, J. E., Nitrate reduction by roots of soybean (*Glycine max* [L.] Merr.) seedlings, *Plant Physiol.,* 69, 1298, 1982.
22. Finke, R. L., Harper, J. E., and Hageman, R. H., Efficiency of nitrogen assimilation by N_2-fixing and nitrate-grown soybean (*Glycine max* [L.] Merr.), *Plant Physiol.,* 70, 1178, 1982.
23. Aslam, M. and Huffaker, R. C., *In vivo* nitrate reduction in roots and shoots of barley (*Hordeum vulgare* L.) seedlings in light and dark, *Plant Physiol.,* 70, 1009, 1982.
24. Radin, J. W., Contribution of the root system to nitrate assimilation in whole cotton plants, *Aust. J. Plant Physiol.,* 4, 811, 1977.
25. Redenbaugh, M. G. and Campbell, W. H., Purification and characterization of NAD(P)H:nitrate reductase and NADH:nitrate reductase from corn roots, *Plant Physiol.,* 68, 115, 1981.
26. Stewart, G. R. and Orebamjo, T. O., Some unusual characteristics of nitrate reduction in *Erythrina senegalensis* DC., *New Phytol.,* 83, 311, 1979.
27. Campbell, W. H. and Black, C. C., Jr., Cellular aspects of C_4 leaf metabolism, in *Cellular and Subcellular Localization in Plant Metabolism,* Creasy, L. and Hrazdina, G., Eds., Plenum Press, New York, 1982, 223.
28. Campbell, W. H., Isolation of NAD(P)H:nitrate reductase from the scutellum of maize, *Z. Pflanzenphysiol.,* 88, 357, 1978.
28a. Oji, Y., Miki, Y., and Okamoto, S., Extraction and affinity purification of NADH:nitrate reductase from barley (*Hordeum distichum* L.) roots, *Plant Cell Physiol.,* 23, 1025, 1982.
29. Dailey, F. A., Warner, R. L., Somers, D. A., and Kleinhofs, A., Characteristics of a nitrate reductase barley mutant deficient in NADH:nitrate reductase, *Plant Physiol.,* 69, 1200, 1982.
30. Cove, D., personal communication, 1982.
31. Garrett, R. H. and Amy, N. K., Nitrate assimilation in fungi, *Adv. Microbiol. Physiol.,* 18, 1, 1978.
32. Smarrelli, J., Jr. and Campbell, W. H., Immunological approaches to structural comparisons of assimilatory nitrate reductases, *Plant Physiol.,* 68, 1226, 1981.
33. Hageman, R. H. and Reed, A. J., Nitrate reductase from higher plants, *Methods Enzymol.,* 69, 270, 1980.
34. Chantarotwong, W., Huffaker, R. C., Miller, B. L., and Granstedt, R. C., *In vivo* nitrate reduction in relation to nitrate uptake, nitrate content, and *in vitro* nitrate reductase activity in intact barley seedlings, *Plant Physiol.,* 57, 519, 1976.
35. Aslam, M., Huffaker, R. C., Rains, D. W., and Rao, K. P., Influence of light and ambient carbon dioxide concentration on nitrate assimilation by intact barley seedlings, *Plant Physiol.,* 63, 1205, 1979.
36. Heuer, B. and Plaut, Z., Reassessment of the *in vivo* assay for nitrate reductase in leaves, *Physiol. Plant.,* 43, 306, 1978.
37. Lawrence, J. M. and Herrick, H. E., Media for in vivo nitrate reductase assay of plant tissues, *Plant Sci. Lett.,* 24, 17, 1982.
38. Aslam, M., Reevaluation of anaerobic nitrite production as an index for the measurement of metabolic pool of nitrate, *Plant Physiol.,* 68, 305, 1981.
39. Scholl, R. L., Harper, J. E., and Hageman, R. H., Improvements in nitrite color development assays of nitrate reductase by phenazine methyl sulfate and zinc acetate, *Plant Physiol.,* 53, 825, 1976.

40. Afridi, M. M. R. K. and Hewitt, E. J., The inducible formation and stability of nitrate reductase in higher plants. I. Effects of nitrate and molybdenum on enzyme activity in cauliflower (*Brassica oleracea* var. botrytis,), *J. Exp. Bot.,* 15, 251, 1964.
41. Smarrelli, J., Jr. and Campbell, W. H., NADH dehydrogenase activity of higher plant nitrate reductase (NADH), *Plant Sci. Lett.,* 16, 139, 1979.
42. Cresswell, C. F., Hageman, R. H., Hewitt, E. J., and Hucklesby, D. P., The reduction of nitrate, nitrite and hydroxylamine to ammonia by enzymes from *Cucurbita pepo* L. in the presence of reduced benzyl viologen as electron donor, *Biochem. J.,* 94, 40, 1965.
43. Jolly, S. O., Campbell, W. H., and Tolbert, N. E., NADPH- and NADH-nitrate reductase from soybean leaves, *Arch. Biochem. Biophys.,* 174, 431, 1976.
44. Orihuel-Iranzo, B. and Campbell, W. H., Development of NAD(P)H: and NADH:nitrate reductase activities in soybean cotyledons, *Plant Physiol.,* 65, 595, 1980.
45. Shen, T., Funkhouser, E. A., and Guerrero, M. G., NADH- and NADPH-nitrate reductase in rice seedlings, *Plant Physiol.,* 58, 292, 1976.
46. Redinbaugh, M. G. and Campbell, W. H., Purification of squash NADH:nitrate reductase by zinc chelate affinity chromatography, *Plant Physiol.,* 71, 205, 1983.
47. Campbell, W. H., Redinbaugh, M. G., Mahony, W. B., and Smarrelli, J., Jr., New forms and new characteristics of higher plant nitrate reductases, in *Photosynthesis IV. Regulation of Carbon Metabolism,* Akoyunoglou, G., Ed., Balaban Int. Sci. Services, Philadelphia, 1981, 707.
48. Mahony, W. B., Purification and Molecular Weight Studies of Higher Plant Nitrate Reductase, M.S. thesis, State University of New York, College of Environmental Science and Forestry, Syracuse, 1982.
49. Hageman, R. H. and Hucklesby, D. P., Nitrate reductase from Higher Plants, *Methods Enzymol.,* 23, 491, 1971.
50. Smarrelli, J., Jr. and Campbell, W. H., Heavy metal inactivation and chelator stimulation of higher plant nitrate reductase, *Biochim. Biophys. Acta,* 742, 435, 1983.
51. Kuo, T., Warner, R. L., and Kleinhofs, A., *In vitro* stability of nitrate reductase from barley leaves, *Phytochemistry,* 21, 531, 1982.
52. Smarrelli, J., Jr., A Study of Higher Plant Nitrate Reductase, M.S. thesis, State University of New York, College of Environmental Science and Forestry, Syracuse, 1977.
53. Smarrelli, J., Jr., Biochemical and Immunological Studies of Assimilatory Nitrate Reductases, Ph.D. thesis, State University of New York, College of Environmental Science and Forestry, Syracuse, 1980.
54. Schrader, L. E., Cataldo, D. A., and Peterson, D. M., Use of protein in extraction and stabilization of nitrate reductase, *Plant Physiol.,* 53, 688, 1974.
55. Wray, J. L. and Kirk, D. W., Inhibition of NADH-nitrate reductase degradation in barley leaf extracts by leupeptin, *Plant Sci. Lett.,* 23, 207, 1981.
56. Kuo, T. M., Somers, D. A., Kleinhofs, A., and Warner, R. L., NADH-nitrate reductase in barley leaves. Identification and amino acid composition of subunit protein, *Biochim. Biophys. Acta,* 708, 75, 1982.
57. Notton, B. A. and Hewitt, E. J., Structure and properties of higher plant nitrate reductase, especially *Spinacea oleracea, in Nitrogen Assimilation of Plants,* Hewitt, E. J. and Cutting, C. V., Eds., Academic Press, London, 1979, 227.
58. Seigel, L. M. and Monty, K. J., Determination of molecular weights and frictional ratios of proteins in impure systems by use of gel-filtration and density gradient centrifugation. Application to crude preparations of sulfite and hydroxylamine reductases, *Biochim. Biophys. Acta,* 112, 346, 1966.
59. Hedrick, J. L. and Smith, A. J., Size and charge isomer separation and estimation of molecular weights of proteins by disc gel electrophoresis, *Arch. Biochem. Biophys.,* 126, 155, 1968.
60. Nakagawa, H. and Oaks, A., Purification of nitrate reductase (NR) from corn leaves (*Zea mays* cr. W64A X W182E), *Plant Physiol.,* 69S, 111, 1982.
61. Roustan, J. -L., Neuburger, M., and Fourcy, A., Nitrate reductase of maize leaves, *Physiol. Veg.,* 12, 527, 1974.
62. Anaker, W. F. and Stoy, V., Proteinchromatographie an calcumphosphat. I. Reinigung von nitratreductase aus Weizenblattern, *Biochem. Zeitschrift.,* 330, 141, 1959.
63. Howard, W. D. and Solomonson, L. P., Quaternary structure of assimilatory NADH:nitrate reductase from *Chlorella, J. Biol. Chem.,* 257, 10243, 1982.
64. De la Rosa, M. A., Vega, J. M., and Zumft, W. G., Composition and structure of assimilatory nitrate reductase from *Ankistrodesmus braunii, J. Biol. Chem.,* 256, 5814, 1981.
65. Pan, S.-S. and Nason, A., Purification and characterization of homogeneous assimilatory reduced nicotinamide adenine dinucleotide phosphate-nitrate reductase from *Neurospora crassa, Biochim. Biophys. Acta,* 523, 297, 1978.
66. Renosto, F., Ornitz, D. M., Peterson, D., and Segel, I. H., Nitrate reductase from *Penicillium chrysogenum,* Purification and kinetic mechanism, *J. Biol. Chem.,* 256, 8616, 1981.

67. Guerrero, M. G. and Gutierrez, M., Purification and properties of the NAD(P)H:nitrate reductase of the yeast *Rhoodotorula glutinis, Biochim. Biophys. Acta,* 482, 272, 1977.

68. Small, I. S. and Wray, J. L., Breakdown of barley NADH-nitrate reductase to functional NADH-cytochrome c reductase species, *Biochem. Soc. Trans.,* 7, 737, 1979.

69. Wray, J. L., Small, I. S., and Brown, J., A model for the subunit composition of higher plant NADH-nitrate reductase, *Biochem. Soc. Trans.,* 7, 739, 1979.

70. Redinbaugh, M. G., Mahony, W. B., and Campbell, W. H., Purification and the molecular weight of nitrate reductase, *Plant Physiol.,* 69S, 116, 1982.

71. Small, I. S. and Wray, J. L., NADH nitrate reductase and related NADH cytochrome c reductase species in barley, *Phytochemistry,* 19, 387, 1980.

72. Brown, J., Small, I. S., and Wray, J. L., Age-dependent conversion of nitrate reductase to cytochrome c reductase species in barley leaf extracts, *Phytochemistry,* 20, 389, 1981.

73. Campbell, J. M. and Wray, J. L., Purification of barley nitrate reductase and demonstration of nicked subunits, presented at Int. Symp. Nitrate Assimilation — Molecular and Genetic Aspects, Gatersleben, G.D.R, June 22 to 24, 1982, 31.

74. Redinbaugh, M. G. and Campbell, W. H., unpublished data, 1983.

75. Wallace, W., Proteolytic inactivation of enzymes, in *Regulation of Enzyme Synthesis and Activity in Higher Plants,* Smith, H., Ed., Academic Press, London, 1977, 177.

76. Jacob, G. S. and Orme-Johnson, W. H., Prosthetic groups and mechanism of action of nitrate reductase from *Neurospora crassa, in Molybdenum and Molybdenum-Containing Enzymes,* Coughlan, M. P., Ed., Pergamon Press, Oxford, 1980, 327.

77. Garrett, R. H. and Nason, A., Involvement of a b-type cytochrome in the assimilatory nitrate reductase of *Neurospora crassa, Proc. Natl. Acad. Sci. U.S.A.,* 58, 1603, 1967.

78. Somers, D. A., Kuo, T., Kleinhofs, A., and Warner, R. L., Barley nitrate reductase contains a functional cytochrome b_{557}, *Plant Sci. Lett.,* 24, 261, 1982.

79. Fido, R. J., Hewitt, E. J., Notton, B. A., Jones, O. T. G., and Nasrulhaq-Boyce, A., Haem of spinach nitrate reductase: low temperature spectrum and mid-point potential, *FEBS Lett.,* 99, 180, 1979.

80. Notton, B. A. and Hewitt, E. J., Incorporation of radioactive molybdenum into protein during nitrate reductase formation and effect of molybdenum on nitrate reductase and diaphorase activities of spinach (*Spinach oleracea* L.), *Plant Cell Physiol.,* 12, 465, 1971.

81. Notton, B. A., Fido, R. J., Watson, E. F., and Hewitt, E. J., Presence of haem in the tungsten analogue of nitrate reductase and its relationship to dehydrogenase function, *Plant Sci. Lett.,* 14, 85, 1979.

82. Johnson, J. L., Hainline, B. E., and Rajagopalan, K. V., Characterization of the molybdenum cofactor of sulfite oxidase, xanthine oxidase, and nitrate reductase, *J. Biol. Chem.,* 255, 1783, 1980.

83. Johnson, J. L., The molybdenum cofactor common to nitrate reductase, xanthine dehydrogenase and sulfite oxidase, in *Molybdenum and Molybdenum-Containing Enzymes,* Coughlan, M. P., Ed., Pergamon Press, Oxford, 1980, 345.

84. Rajagopalan, K. V., Johnson, J. L., and Hainline, B. E., The pterin of the molybdenum cofactor, *Fed. Proc. Fed. Am. Soc. Exp. Biol.,* 41, 2608, 1982.

85. Johnson, J. L. and Rajagopalan, K. V., Structural and metabolic relationship between the molybdenum cofactor and urothione, *Proc. Natl. Acad. Sci. U.S.A.,* 79, 6856, 1982.

86. Alikulov, Z. A., Lvov, N. P., Burikhanov, Sh. S., and Kretovich, V. L., Isolation of cofactor from molybo-enzymes: nitrate reductase of lupin bacteroids and xanthinoxidase of milk (in Russian), *Izvet. Acad. Sci. U.S.S.R.,* Ser. Biol., 5, 712, 1980.

87. Mendel, R. R., Alikulov, Z. A., Lvov, N. P., and Muller, A. J., Presence of the molybdenum-cofactor in nitrate reductase-deficient mutant cell lines of *Nicotiana tabacum, Mol. Gen. Genet.,* 181, 395, 1981.

88. Mendel, R. R., Alikulov, Z. A., and Muller, A. J., Molybdenum cofactor in nitrate reductase-deficient tobacco mutants. II. Release of cofactor by heat treatment, *Plant Sci. Lett.,* 25, 67, 1982.

89. Mendel, R. R., Alikulov, Z. A., and Muller, A. J., Molybdenum cofactor in nitrate reductase-deficient tobacco mutants. III. Induction of cofactor synthesis by nitrate, *Plant Sci. Lett.,* 27, 95, 1982.

90. Wallace, W. and Johnson, C. B., Nitrate reductase and soluble cytochrome c reductases in higher plants, *Plant Physiol.,* 61, 748, 1978.

91. Schulz, G. E., Schirmer, R. H., Sachsenheimer, W., and Pai, E. F., The structure of the flavoenzyme glutathione reductase, *Nature (London),* 273, 120, 1978.

92. Richardson, J. S., The anatomy and taxonomy of protein structure, *Adv. Protein Chem.,* 34, 167, 1981.

93. Palmer, G. and Olson, J. S., Concepts and approaches to the understanding of electron transfer processes in enzymes containing multiple redox centers, in *Molybdenum and Molybdenum-Containing Enzymes,* Coughlan, M. P., Ed., Pergamon Press, Oxford, 1980, 187.

94. Couglan, M. P., Ed., *Molybdenum and Molybdenum-Containing Enzymes,* Pergamon Press, Oxford, 1980.
95. Newton, W. E. and Otsuka, S., Eds., *Molybdenum Chemistry of Biological Significance,* Plenum Press, New York, 1980.
96. Campbell, W. H., Affinity purification and kinetic studies of squash cotyledon NADH nitrate reductase, in *Nitrogen Assimilation of Plants,* Hewitt, E. J. and Cutting, C. V., Eds., Academic Press, London, 1979, 321.
97. Smarrelli, J., Jr. and Campbell, W. H., Iodate reduction catalyzed by marine algal nitrate reductase, *Plant Physiol.,* 61S, 53, 1978.
98. Tsunogai, S. and Sase, T., Formation of iodide-iodine in the ocean, *Deep-Sea Res.,* 15, 489, 1969.
99. Oostindier-Braaksma, F. J. and Feenstra, W. J., Isolation and characterization of chlorate-resistant mutants of *Arabidopsis thaliana, Mutation Res.,* 19, 175, 1973.
100. Jawali, N., Ramakrishna, J., Sainis, J. K., and Sane, P. V., Inhibition of the nitrate reductase complex by dibromothymoquinone, *Z. Naturforsch.,* 34c, 529, 1979.
101. De la Rosa, F. F., Castillo, F., and Palacian, E., Effects of denaturing agents on spinach nitrate reductase, *Phytochemistry,* 16, 875, 1977.
102. Schrader, L. E., Ritenour, G. L., Eilrich, G. L., and Hageman, R. H., Some characteristics of nitrate reductase from higher plants, *Plant Physiol.,* 43, 930, 1968.
103. Orihuel-Iranzo, B. and Campbell, W. H., Pyridine nucleotide specificity and kinetics of soybean leaf NAD(P)H: nitrate reductase , in *Photosynthesis IV. Regulation of Carbon Metabolism,* Akoyunoglou, G., Ed., Balaban Int. Sci. Services, Philadelphia, 1981, 715.
104. Jawali, N., Sainis, J. K., and Sane, P. V., Hydroxylamine inhibition of the nitrate reductase complex from *Amaranthus, Phytochemistry,* 17, 1527, 1978.
105. Olsen, R. A., Clark, R. B., and Bennett, J. H., The enhancement of soil fertility by plant roots, *Am. Sci.,* 69, 378, 1981.
106. Howard, W. D. and Solomonson, L. P., Kinetic mechanism of assimilatory NADH:nitrate reductase from *Chlorella, J. Biol. Chem.,* 256, 12725, 1981.
107. McDonald, D. W. and Coddington, A., Properties of the assimilatory nitrate reductase from *Aspergillus nidulans, Eur. J. Biochem.,* 46, 169, 1974.
108. Renosto, F., Schmidt, N. D., and Segel, I. H., Nitrate reductase from *Pennicillium chrysogenum:* kinetic mechanism at sub-optimum pH, *Biochem. Biophys. Res. Commun.,* 107, 12, 1982.
109. De la Rosa, F. F., Palacian, E., and Castillo, F., Studies on the kinetic mechanism of nitrate reductase from spinach (*Spinacea oleracea), Rev. Esp. Fisiol.,* 36, 279, 1980.
110. Herrero, A., De la Rosa, M. G., Diez, J., and Vega, J. M., Catalytic properties of *Ankistrodesmus braunii* nitrate reductase, *Plant Sci. Lett.,* 17, 409, 1980.
111. Coughlan, M. P. and Rajagopalan, K. V., The kinetic mechanism of xanthine dehydrogenase and related enzymes, *Eur. J. Biochem.,* 105, 81, 1980.
112. Orihuel, B., Studies of Soybean Nitrate Reductase, M.S. thesis, State University of New York, College of Environmental Science and Forestry, Syracuse, 1980.
113. Dailey, F. A., Kuo, T., and Warner, R. L., Pyridine nucleotide specificity of barley nitrate reductase, *Plant Physiol.,* 69, 1196, 1982.
114. Rendina, A. R. and Orme-Johnson, W. H., Glutamate synthase: on the kinetic mechanism of the enzyme from *Escherichia coli* W, *Biochemistry,* 17, 5388, 1978.
115. Beevers, L., Flesher, D., and Hageman, R. H., Studies on the pyridine nucleotide specificity of nitrate reductase in higher plants and its relationship to sulfhydryl level, *Biochim. Biophys. Acta,* 89, 453, 1964.
116. Wells, G. N. and Hageman, R. H., Specificity for nicotinamide dinucleotide by nitrate reductase from leaves, *Plant Physiol.,* 54, 136, 1974.
117. Jolly, S. O., Nitrate Reductase and a Nitrate Reductase Inhibitor in Soybean Leaves, Ph.D. thesis, Michigan State University, East Lansing, 1975.
118. Graf, L., Notton, B. A., and Hewitt, E. J., Serological estimation of spinach nitrate reductase, *Phytochemistry,* 14, 1241, 1975.
119. Notton, B. A., Graf, L., Hewitt, E. J., and Povey, R. C., The role of molybdenum in the synthesis of nitrate reductase in cauliflower (*Brassica oleracea* L. var. Botrytis L.) and Spinach (*Spinacea oleracea* L.), *Biochim. Biophys. Acta,* 364, 45, 1974.
120. Pateman, J. A., Cove, D. J., Rever, B. M., and Roberts, D. B., A common co-factor for nitrate reductase and xanthine dehydrogenase which also regulates the synthesis of nitrate reductase, *Nature (London),* 201, 58, 1964.
121. Amy, N. K. and Garrett, R. H., Immunoelectrophoretic determination of nitrate reductase in *Neurospora crassa, Anal. Biochem.,* 95, 97, 1979.
122. Funkhouser, E. A. and Ramadoss, C. S., Synthesis of nitrate reductase in *Chlorella.* II. Evidence for synthesis in ammonia-grown cells, *Plant Physiol.,* 65, 944, 1980.

123. Kuo, T., Kleinhofs, A., Somers, D., and Warner, R. L., Antigenicity of nitrate reductase-deficient mutants in *Hordeum vulgare* L., *Mol. Gen. Genet.*, 181, 20, 1981.
123a. Somers, D. A., Kuo, T. M., Kleinhofs, A., and Warner, R. L., Nitrate reductase-deficient mutants in barley. Immunoelectrophoretic characterization, *Plant Physiol.*, 71, 145, 1983.
124. Ripp, K. G. and Campbell, W. H., unpublished data, 1983.
125. Schiemann, J., personal communication, 1982.
126. Somers, D. A., Kuo, T., Warner, R. L., Kleinhofs, A., and Oaks, A., Regulation of nitrate reductase in barley seedlings, *Plant Physiol.*, 69S, 128, 1982.
127. Stewart, G. R. and Rhodes, D., Control of enzyme levels in the regulation of nitrogen metabolism, in *Regulation of Enzyme Synthesis and Activity in Higher Plants,* Smith, H., Ed., Academic Press, London, 1977, 1.
127a. Gupta, A., Saxena, I. M., Sopory, S. K., and Guha-Mukherjee, S., Regulation of ntirate reductase synthesis during early germination in seeds of barley *(Hordeum vulgare)*, *J. Exp. Bot.*, 34, 34, 1983.
128. Ingle, J., Nucleic acid and protein synthesis associated with the induction of nitrate reductase activity in radish cotyledons, *Biochem. J.*, 108, 715, 1968.
129. Beevers, L. and Hageman, R. H., The role of light in nitrate metabolsim in higher plants, in *Current Topics in Photobiology and Photochemistry,* Vol. VII, Giese, A. C., Ed., Academic Press, New York, 1972, 85.
129a. Huffaker, R. C., Biochemistry and physiology of leaf proteins, in *Nucleic Acids and Proteins in Plants. I. Structure, Biochemistry and Physiology of Proteins,* Encyclopedia of Plant Physiology, New Series Vol. 14A, Boulter, D. and Parthier, B., Eds., Springer-Verlag, Berlin, 1982, chap. 10.
130. Zielke, H. R. and Filner, P., Synthesis and turnover of nitrate reductase induced by nitrate in cultured tobacco cells, *J. Biol. Chem.*, 246, 1772, 1971.
131. Radin, J. W., Differential regulation of nitrate reductase induction in roots and shoots of cotton plants, *Plant Physiol.*, 55, 178, 1975.
132. Sihag, R. K., Guha-Mukherjee, S., and Sopory, S. K., Regulation of nitrate reductase level in pea: *in vivo* stability by ammonium, *Biochem. Biophys. Res. Commun.*, 85, 1017, 1978.
133. Sihag, R. K., Guha-Mukherjee, S., and Sopory, S. K., Effect of ammonium, sucrose and light on the regulation of nitrate reductase level in *Pisum sativum*, *Physiol. Plant.*, 45, 281, 1979.
134. Srivastava, H. S., Asthana, J. S., and Jain, A., Induction of nitrate reductase and peroxidase activity in the seedlings of normal and high lysine maize, *Physiol. Plant.*, 47, 199, 1979.
135. Mehta, P. and Srivastava, H. S., Comparative stability of ammonium and nitrate induced nitrate reductase activity in maize leaves, *Phytochemistry*, 19, 2527, 1980.
136. Mehta, P. and Srivastava, H. S., Increase in *in vivo* nitrate reductase activity in maize leaves in the presence of ammonium, *Biochem. Physiol. Pflanzen.*, 177, 237, 1982.
137. Mackown, C. T., Jackson, W. A., and Volk, R. J., Restricted nitrate influx and reduction in corn seedlings exposed to ammonium, *Plant Physiol.*, 69, 353, 1982.
138. Knypl, J. S., Hormonal control of nitrate assimilation: do phytochromones and phytochrome control the activity of nitrate reductase?, in *Nitrogen Assimilation of Plants,* Hewitt, E. J. and Cutting, C. V., Eds., Academic Press, London, 1979, 541.
139. Street, H. E., Plant cell cultures: their potential for metabolic studies, in *Biosynthesis and its Control in Plants,* Milborrow, B. V., Ed., Academic Press, London, 1972, 93.
140. Campbell, W. H., Ziegler, P., and Beck, E., unpublished data, 1983.
141. Husemann, W. and Barz, W., Photoautotrophic growth and photosynthesis in cell suspension cultures of *Chenopodium rubrum*, *Physiol. Plant.*, 40, 77, 1977.
142. Filner, R., Regulation of nitrate reductase in cultured tobacco cells, *Biochim. Biophys. Acta*, 118, 299, 1966.
143. Heimer, Y. A. and Filner, P., Regulation of the nitrate assimilation pathway in cultured tobacco cells. II. Properties of a variant cell line, *Biochim. Biophys. Acta*, 215, 152, 1970.
144. Heimer, Y. A. and Filner, P., Regulation of the nitrate assimilation pathway in cultured tobacco cells. III. The nitrate uptake system, *Biochim. Biophys. Acta*, 230, 362, 1971.
145. Kelker, H. C. and Filner, P., Regulation of nitrate reductase and its relationship to the regulation of nitrate reductase in cultured tobacco cells, *Biochim. Biophys. Acta*, 252, 69, 1971.
146. Heimer, Y. A. and Riklis, E., On the mechanism of development of nitrate reductase activity in tobacco cells, *Plant Sci. Lett.*, 16, 135, 1979.
147. Mohanty, B. and Fletcher, J. S., Ammonium influence on the growth and nitrate reductase activity of Paul's Scarlet rose suspension cultures, *Plant Physiol.*, 58, 152, 1976.
148. Mohanty, B. and Fletcher, J. S., Influence of ammonium on the growth and development of suspension cultures of Paul's Scarlet rose, *Physiol. Plant.*, 42, 221, 1978.
149. Jordan, D. B. and Fletcher, J. S., The relationship between NO_2^- accumulation, nitrate reductase and nitrite reductase in suspension cultures of Paul's Scarlet rose, *Plant Sci. Lett.*, 17, 95, 1979.
150. Mohanty, B. and Fletcher, J. S., Ammonium influence on nitrogen assimilating enzymes and protein accumulation in suspension cultures of Paul's Scarlet rose, *Physiol. Plant.*, 48, 453, 1980.

151. Bradford, J. A. and Fletcher, J. S., Influence of protein systhesis on NO_3^- reduction NH_4^+ accumulation, and amide synthesis in suspension cultures of Paul's Scarlet rose, *Plant Physiol.,* 69, 63, 1982.

152. Pandy, K. N. and Sabharwal, P. S., Gamma-irradiation activates biochemical systems: induction of nitrate reductase activity in plant callus, *Proc. Natl. Acad. Sci. U.S.A.,* 79, 5460, 1982.

153. Jones, R. W., Abbott, A. J., Hewitt, E. J., James, D. M., and Best, G. R., Nitrate reductase activity and growth in Paul's Scarlet rose suspension cultures in relation to nitrogen source and molybdenum, *Planta,* 133, 27, 1976.

154. Jones, R. W., Abbott, A. J., Hewitt, E. J., Best, G. R., and Watson, E. F., Nitrate reductase activity in Paul's Scarlet rose suspension cultures and the differential role of nitrate and molybdenum in induction, *Planta,* 141, 183, 1978.

155. Gamborg, O. L., The effects of amino acids and ammonium on the growth of plant cells in suspension culture, *Plant Physiol.,* 45, 372, 1970.

156. Bayley, J. M., King, J., and Gamborg, O. L., The effect of the source of inorganic nitrogen on growth and enzymes of nitrogen assimilation in soybean and wheat cells in suspension cultures, *Planta,* 105, 15, 1972.

157. Oaks, A., The regulation of nitrate reductase in suspension cultures of soybean cells, *Biochim. Biophys. Acta,* 372, 122, 1974.

158. Doddema, H., Hofstra, J. J., and Feenstra, W. J., Uptake of nitrate by mutants of *Arabidopsis thaliana,* disturbed in uptake or reduction of nitrate. I. Effect of nitrogen source during growth on uptake of nitrate and chlorate, *Physiol. Plant.,* 43, 343, 1978.

159. King, J. and Khana, V., A nitrate reductase-less variant isolated from suspension cultures of *Datura innoxia* (Mill.), *Plant Physiol.,* 66, 632, 1980.

160. Murphy, T. M. and Imbrie, C. W., Induction and characterization of chlorate-resistant strains of *Rosa damascena* cultured cells, *Plant Physiol.,* 67, 910, 1981.

161. Strauss, A., Bucher, F., and King, P. J., Isolation of biochemical mutants using haploid mesophyll protosplasts of *Hyoscyamus muticus.* I. A NO_3^- non-utilizing clone, *Planta,* 153, 75, 1981.

162. Braaksma, F. J. and Feenstra, W. J., Nitrate reduction in the wildtype and a nitrate reductase deficient mutant of *Arabidopsis thaliana, Physiol. Plant.,* 54, 351, 1982.

163. Muller, A. J., Genetic studies on nitrate reduction in *Nicotiana* and *Hordeum vulgare,* presented at Int. Symp. Nitrate Assimilation — Molecular and Genetic Aspects, Gatersleben, G.D.R, June 22 to 24, 1982, 9.

164. Marton, L., Dung, T. M., Mendel, R. R., and Maliga, P., Nitrate reductase deficient cell lines from haploid protoplast cultures of *Nicotiana plumbaginifolia, Mol. Gen. Genet.,* 182, 301, 1982.

165. Evola, S. V., Chlorate-resistant variants of *Nicotiana tabacum* L. I. Selection *in vitro* and phenotypic characterization of cell lines and regenerated plants, in preparation, 1983.

166. Muller, A. J. and Grafe, R., Isolation and characterization of cell lines of *Nicotiana tabacum* lacking nitrate reductase, *Mol. Gen. Genet.,* 161, 67, 1978.

167. Braaksma, F. J., Genetic Control of Nitrate Reduction in *Arabidopsis thaliana,* Ph.D. thesis, University of Groningen, Haren, The Netherlands, 1982.

168. Feenstra, W. J. and Braaksma, F. J., Reverse mutants of the nitrate reductase-deficient mutant B-25 of *Arabidopsis thaliana,* presented at Int. Symp. Nitrate Assimilation — Molecular and Genetic Aspects, Gatersleben, G.D.R., June 22 to 24, 1982.

169. Mendel, R. R. and Muller, A. J., Reconstitution of NADH-nitrate reductase in vitro from nitrate reductase-deficient *Nicotiana tabacum* mutants, *Mol. Gen. Genet.,* 161, 77, 1978.

170. Mendel, R. R. and Muller, A. J., Nitrate reductase-deficient mutant cell lines of *Nicotiana tabacum.* Further biochemical characterization, *Mol. Gen. Genet.,* 177, 145, 1979.

171. Glimelius, K., Eriksson, T., Grafe, R., and Muller, A. J., Somatic hybridization of nitrate reductase-deficient mutants of *Nicotiana tabacum* by protoplast fusion, *Physiol. Plant.,* 44, 273, 1978.

172. Muller, A. J., NADH-nitrate reductase of *Nicotiana tabacum* is coded for by duplicate structural genes, presented at Int. Symp. Nitrate Assimilation — Molecular and Genetic Aspects, Gatersleben, G.D.R, June 22 to 24, 1982, 23.

173. Warner, R. L. and Kleinhofs, A., Nitrate reductase-deficient mutants in barley, *Nature (London),* 269, 406, 1977.

174. Warner, R. L. and Kleinhofs, A., Nitrate utilization by nitrate reductase-deficient barley mutants, *Plant Physiol.,* 67, 740, 1981.

175. Somers, D. A., Kuo, T., Kleinhofs, A., and Warner, R. L., Immunoelectrophoretic characterization of nitrate reductase deficient barley mutants, *Plant Physiol.,* 67S, 136, 1981.

176. Cove, D. J., Genetic studies of nitrate assimilation in *Aspergillus nidulans, Biol. Rev.,* 54, 291, 1979.

177. Gallagher, L. W., Soliman, K. M., Qualset, C. O., Huffaker, R. C., and Rains, D. W., Major gene control of nitrate reductase activity in common wheat, *Crop Sci.,* 20, 717, 1980.

178. Cove, D. J. and Pateman, J. A., Autoregulation of the synthesis of nitrate reductase in *Aspergillus nidulans, J. Bacteriol.,* 97, 1374, 1969.

179. Mendel, R. R., Molybdenum cofactor and molybdenum cofactor-mutants in tobacco, presented at Int. Symp. Nitrate Assimilation — Molecular and Genetic Aspects, Gatersleben, G.D.R., June 22 to 24, 1982, 11.

180. Yamaya, T., Oaks, A., and Boesel, I. L., Characteristics of nitrate reductase-inactivating proteins obtained from corn roots and rice cell cultures, *Plant Physiol.,* 65, 141, 1980.

181. Wallace, W. and Shannon, J. D., Proteolytic activity and nitrate reductase inactivation in maize seedlings, *Aust. J. Plant Physiol.,* 8, 211, 1981.

182. Yamaya, T., Solomonson, L. P., and Oaks, A., Action of corn and rice inactivating proteins on a purified nitrate reductase from *Chlorella, Plant Physiol.,* 65, 146, 1980.

183. Sherrard, J. H., Kennedy, J. A., and Dalling, M. J., *In vitro* stability of nitrate reductase from wheat leaves. III. Isolation and partial characterization of a nitrate reductase inactivating factor, *Plant Physiol.,* 64, 640, 1979.

184. Yamaya, T. and Ohira, K., Nitrate reductase inactivating factor from rice cells in suspension culture, *Plant Cell Physiol.,* 17, 633, 1976.

185. Yamaya, T. and Ohira, K., Purification and properties of a nitrate reductase inactivating factor from rice cells in suspension culture, *Plant Cell Physiol.,* 18, 915, 1977.

186. Yamaya, T. and Ohira, K., Nitrate reductase inactivating factor from rice seedlings, *Plant Cell Physiol.,* 19, 211, 1978.

187. Yamaya, T. and Ohira, K., Reversible inactivation of nitrate reductase by its inactivating factor from rice cells in suspension culture, *Plant Cell Physiol.,* 19, 1085, 1978.

188. Leong, C. C. and Shen, T.-C., Nitrate reductase inhibitor of rice plants, *Biochim. Biophys. Acta,* 612, 245, 1980.

189. Leong, C. C. and Shen, T.-C., Occurrence of nitrate reductase inhibitor in rice plants, *Plant Physiol.,* 70, 1762, 1982.

190. Leong, C. C. and Shen, T.-C., Action kinetics of the inhibition of nitrate reductase of rice plants, *Biochim. Biophys. Acta,* 703, 129, 1983.

191. Jolly, S. O. and Tolbert, N. E., NADH-nitrate reductase inhibitor from soybean leaves, *Plant Physiol.,* 62, 197, 1978.

192. Kadam, S. S., Gandhi, A. P., Sawhney, S. K., and Naik, M. S., Inhibitor of nitrate reductase in the roots of rice seedlings and its effect on the enzyme activity in the presence of NADH, *Biochim. Biophys. Acta,* 350, 162, 1974.

193. Kadam, S. S., Sawhney, S. K., and Naik, M. S., Effect of NADH on the activities of nitrate reductase and its inactivating enzyme, *Indian J. Biochem. Biophys.,* 12, 81, 1975.

194. Vennesland, B., HCN and the control of nitrate reduction. The regulation of the amount of active nitrate reductase present in chlorella cells, in *Biology of Inorganic Nitrogen and Sulfur,* Bothe, H. and Trebst, A., Eds., Springer-Verlag, Berlin, 1981, 233.

195. Smarrelli, J., Jr. and Campbell, W. H., Activation of *Thalassiosira pseudonana* NADH: nitrate reductase, *Phytochemistry,* 19, 1601, 1980.

196. Sherrard, J. H., Kennedy, J. A., and Dalling, M. J., *In vitro* stability of nitrate reductase from wheat leaves. II. Isolation of factors from crude extract which affect stability of highly purified nitrate reductase inactivating factor, *Plant Physiol.,* 64, 439, 1979.

197. Yamaya, T. and Oaks, A., Activation of nitrate reductase by extracts from corn scutella, *Plant Physiol.,* 66, 212, 1980.

198. Ekes, M., Ultrastructural demonstration of ferricyanide reductase (diaphorase) activity in the envelopes of the plastids of etiolated barley (*Hordeum vulgare* L.) leaves, *Planta,* 151, 439, 1981.

199. Vaughn, K. C. and Duke, S. O., Histochemical localization of nitrate reductase, *Histochemistry,* 72, 191, 1981.

200. Roldan, J. M., Verbelen, J.-P., Butler, W. L., and Tokuyasu, K., Intracellular localization of nitrate reductase in *Neurospora crassa, Plant Physiol.,* 70, 872, 1982.

201. Martinova, E., Hock, U., and Wiemken, A., Vacuoles as storage compartments for nitrate in barley leaves, *Nature (London),* 289, 292, 1981.

202. Granstedt, R. C. and Huffaker, R. C., Identification of the leaf vacuole as a major nitrate storage pool, *Plant Physiol.,* 70, 410, 1982.

203. Brown, J. C., Genetically controlled chemical factors involved in absorption and transport of iron by plants, in *Bioinorganic chemistry-II,* Raymond, K. N., Ed., American Chemical Society, Washington, D.C., 1977, 93.

204. Lee, R. B., Sources of reductant for nitrate assimilation in non-photosynthetic tissue: a review, *Plant Cell Environ.,* 3, 65, 1980.

205. Reed, A. J. and Canvin, D. T., Light and dark controls of nitrate reduction in wheat (*Triticum aestivum* L.) protoplasts, *Plant Physiol.,* 69, 508, 1982.

206. Stitt, M., Lilley, R. M., and Heldt, H., Adenine nucleotide levels in the cytosol, chloroplasts, and mitochrondira of wheat protoplasts, *Plant Physiol.,* 70, 971, 1982.

207. Ben-Shalom, N., Huffaker, R. C., and Rappaport, L., Effect of photosynthetic inhibitors and un-couplers of oxidative phosphorylation on nitrate and nitrite reduction in barley leaves, *Plant Physiol.*, 71, 63, 1983.
208. Nasrulhaq-Boyce, A. and Jones, O. T. G., The light-induced development of nitrate reductase in etiolated barley shoots: an inhibitory effect of levulinic acid, *Planta,* 137, 77, 1977.
209. Wolf, D. H., Proteinase action *in vitro* versus proteinase function *in vivo:* mutants shed light on intracellular proteolysis in yeast, *Trends Biochem. Sci.,* 7, 35, 1982.
210. Marzluf, G. A., Regulation of nitrogen metabolism and gene expression in fungi, *Microbiol. Rev.,* 45, 437, 1981.
211. Tomsett, A. B. and Garrett, R. H., Biochemical analysis of mutants defective in nitrate assimilation in *Neurospora crassa:* evidence for autogenous control by nitrate reductase, *Mol. Gen. Genet.,* 184, 183, 1981.
212. Feenstra, W. J. and Jacobsen, E., Isolation of a nitrate reductase deficient mutant of *Pisum sativum* by means of selection for chlorate resistance, *Theor. Appl. Genet.,* 58, 39, 1981.
213. Cocking, E. C., Davey, M. R., Pental, D., and Power, J. B., Aspects of plant genetic manipulation, *Nature (London),* 293, 265, 1981.
214. Ryan, L. D. and Vestling, C. S., Rapid purification of lactate dehydrogenase from rat liver and hepatoma: a new approach, *Arch. Biochem. Biophys.,* 160, 279, 1974.
215. Solomonson, L. P., Purification of NADH-nitrate reductase by affinity chromatography, *Plant Physiol.,* 56, 853, 1975.
216. Guerrero, M. G., Jetschmann, K., and Volker, W., The stereo-specificity of nitrate reductase for hydrogen removal from reduced pyridine nucleotides, *Biochim. Biophys. Acta,* 482, 19, 1977.
217. Campbell, W. H. and Smarrelli, J., Jr., Affinity purification of higher plant nitrate reductase, *Plant Physiol.,* 57S, 37, 1976.
218. Bohem, H. J., Kopperschlager, G., Schulz, J., and Hoffman, E., Affinity chromatography of phosphofructokinase using cibacron blue F3G-A, *J. Chromatogr.,* 69, 209. 1972.
219. Mendel, R. R., Comparative affinity chromatography of nitrate reductase from wild-type and molybdenum cofactor-deficient cell cultures of *Nicotiana tabacum, Biochem. Physiol. Pflanzen,* 175, 216, 1980.
220. Dudman, W. F. and Bishop, C. T., Electrophoresis of dyed polysaccharides on celluluse acetate, *Can. J. Chem.,* 46, 3079, 1968.
221. Porath, J., Carlsson, J., Olsson, I., and Bel Fraye, O., Metal chelate affinity chromatography, a new approach to protein fractionation, *Nature (London),* 258, 598, 1975.
222. Bruck, C., Portetelle, D., Glineur, C., and Bollen, A., One-step purification of mouse monoclonal antibodies from ascitic fluid by DEAE Affi-Gel Blue chromatography, *J. Immunol. Methods,* 53, 313, 1982.
223. Wilson, M. B. and Nakane, P. K., Recent developments in the periodate method of conjugating horseradish peroxidase (HRPO) to antibodies, in *Immunofluourescence and Related Staining Techniques,* Knapp, W., Holubar, K., and Wick, G., Eds., Elsevier/North-Holland, Amsterdam, 1978, 215.
224. Avrameas, S., Ternynck, T., and Guesdon, J.-L., Coupling of enzymes to antibodies and antigens, *Scand. J. Immunol.,* 8 (Suppl. 7), 7, 1978.

Chapter 2

NITROGENASE: PROPERTIES AND REGULATION

Paul W. Ludden and Robert H. Burris

TABLE OF CONTENTS

I. INTRODUCTION

Breeders are interested in developing lines of plants with enhanced ability for biological nitrogen fixation. Water commonly limits the growth of plants, but nitrogen is next to it as the most frequently limiting factor. Nitrogen can be supplied as fertilizer, or in the case of N_2-fixing plants it can be assimilated from the air.

One of our largest chemical industries is concerned with the chemical fixation of nitrogen. The world capacity for chemical fixation is about 60 million metric tons of N per year, and the plants usually operate with an annual output near 50 million metric tons. Of this, nearly 40 million metric tons is utilized for fertilizer. The biological process far surpasses chemical fixation. Although it is difficult to assign a specific amount to biological N_2 fixation, the most credible estimates are in the range of 100 to 180 million metric tons per year.[1] It is apparent that biological fixation is of prime importance in maintaining the nitrogen economy on earth. Losses to the sea occur constantly through leaching and erosion, and N_2 is returned to the atmosphere through denitrification. The terrestrial nitrogen balance must be maintained by replacement through chemical and biological nitrogen fixation.

There is interest in breeding N_2-fixing plants and free-living organsims for enhanced fixation, and the problem can be approached in a number of ways. One could attempt to improve the efficiency of the process or to improve the generation of ATP which serves as a prime source of energy for N_2 fixation. One also might attempt to increase the abundance of nitrogenase or to increase the concentration of ferredoxin, flavodoxin, or other auxiliary electron carriers in the system. As nitrogenase produces H_2, and its dissipation represents a loss of energy, attention has been directed towards enhancing the action of hydrogenases that can recycle the H_2. In the symbiotic system one might attempt to delay senescence of the plant so that it maintains as active N_2 fixation over a longer period. Enhancement of photosynthesis by the plant could generate more reductant to support the nitrogenase system.

II. MECHANISM OF N_2 FIXATION

A. Scheme for Fixation

To understand the various approaches for breeding for N_2 fixation, one must have some appreciation of the mechanism of the process.[2] Figure 1 shows an outline scheme for N_2 fixation, and we will discuss the activity of the components in this system.

B. Components of the Nitrogenase System
1. Dinitrogenase

Nitrogenase is a term used to embrace the complex of two proteins responsible for N_2 fixation. The first protein is dinitrogenase, a MoFe protein. This large protein has a molecular weight[3] that varies with the source but usually is from 200,000 to 234,000. It contains 2 Mo per molecule and about 32 Fe per molecule; acid labile sulfur roughly matches the number of atoms of Fe. The protein is an $\alpha_2\beta_2$ molecule; i.e., it consists of four subunits of two different types. It can be dissociated into these parts, but no one has been able to reassemble the parts in an active form. Dinitrogenase has a very characteristic EPR spectrum[4] that changes with its oxidation-reduction state, and this characteristic has been very helpful in unraveling the electron transport pathway in the nitrogenase system. In the fully reduced state the EPR signal disappears, and it reappears when the dinitrogenase returns to a partially reduced state. There also is an oxidized form that is EPR silent, but there is no evidence that the shift between the fully oxidized and the partially reduced form is physiologically significant.

Catalytic activity for substrate reduction centers in the MoFe portion of the mole-

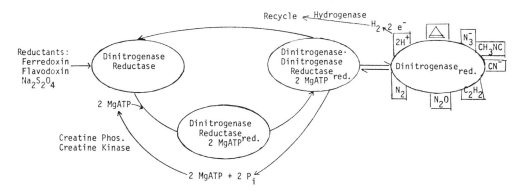

FIGURE 1.   A working hypothesis for electron flow through the nitrogenase enzyme system.

cule. This concept has been supported clearly by the work on FeMoco, an iron-molybdenum cofactor that can be removed from nitrogenase. FeMoco when added to a catalytically inactive extract from a defective mutant of *Azotobacter vinelandii* restores activity to the nitrogenase system.[5] This constitutes strong evidence for the catalytic activity of the FeMo center. Dinitrogenase has been recovered in a homogenous form from a variety of nitrogen-fixing organisms, and from some preparations it has been crystallized. The consensus is that dinitrogenase binds and reduces substrates.

2. Dinitrogenase Reductase

The other protein component of the nitrogenase system is referred to as dinitrogenase reductase or the Fe protein. It has a molecular weight in the neighborhood of 60,000 and consists of two identical subunits.[3] Again, the compound can be dissociated into subunits but not reassembled in an active form. Each protein molecule contains an active center with four Fe and four acid-labile S atoms. The scheme shows that dinitrogenase reductase is the agent that transfers electrons for the reduction of dinitrogenase. Hence, the name dinitrogenase reductase is logical, as there is no known function for the Fe protein other than as an electron transfer agent to dinitrogenase. The dinitrogenase in turn binds the substrates and effects their reduction. The dinitrogenase reductase has a somewhat lower isoelectric point than dinitrogenase, and this serves as a basis for its longer retention and separation from dinitrogenase on a column of DEAE cellulose. Both dinitrogenase reductase and dinitrogenase are labile to O_2, and the O_2 inactivation is irreversible. The dinitrogenase reductase is the more O_2 sensitive of the two proteins. All manipulations with the two proteins must be done anaerobically to avoid inactivation. The EPR spectrum of dinitrogenase reductase is rather characteristic of that of a number of iron proteins, and it is not as unusual as the spectrum of dinitrogenase.[4]

3. ATP

Biological N_2 fixation, like chemical N_2 fixation, is an energy-demanding process. The main source of energy is furnished by MgATP in the biological system. Additional energy is furnished to the system by a reductant. The MgATP initiates its activity by binding specifically to dinitrogenase reductase.[6] Two molecules of MgATP bind per molecule of dinitrogenase reductase.[7] When this binding occurs, there is a marked conformational change in the protein that is accompanied by a decrease of about 100 mV in the oxidation-reduction potential.[8] Normally the dinitrogenase reductase has a potential of about −300 mV, but upon binding the MgATP its potential decreases to a value near −400 mV. This decrease in potential makes it possible for the dinitrogenase reductase to transfer electrons to the dinitrogenase.

There has been some question regarding the number of electrons transferred from dinitrogenase reductase to dinitrogenase. It has been suggested that two electrons are transferred,[9] but our investigations indicate that only one electron is transferred per molecule of dinitrogenase reductase.[10] This transfer of one electron is accompanied by the hydrolysis of two molecules of MgATP,[11] so it is necessary to invest 4 MgATP to transfer a pair of electrons. As the reduction of N_2 to 2 NH_3 is a six-electron process, it is necessary to invest at least 12 MgATP for the reduction of one molecule of N_2. Experiments in several laboratories have indicated that 12 MgATP per N_2 is probably a minimal value, and occasional reports of higher efficiency have not been repeatable. In the laboratory one can arrange the system to give an efficiency of 12 MgATP per N_2, whereas in intact organisms it is common to find a requirement of 20 to 30 MgATP per N_2 reduced.[12] The difference probably results from the fact that experimentally one can provide an optimal balance between the two protein components of the system and can supply an abundant source of reductant to achieve the highest efficiency from the system.

The literature often refers to the nitrogenase complex and suggests that the dinitrogenase and dinitrogenase reductase form a relatively tight functional complex. More convincing evidence suggests that the two proteins form a transitory complex that persists just long enough to transfer one electron; then they dissociate so that the dinitrogenase reductase again can be reduced so that it again can pass another electron to the dinitrogenase.[13] It is possible with certain combinations of nitrogenase components (notably dinitrogenase from *A. vinelandii* and dinitrogenase reductase from *Clostridium pasteurianum*) to construct tight-binding inactive complexes between the two components.[14] Apparently it is obligatory that the two components associate and dissociate rather rapidly to maintain an effective electron flux in the system; tight binding inactivates them.

There has been a question about where ATP is hydrolyzed in the electron transfer scheme. There is rather convincing evidence now that ATP hydrolysis accompanies the electron transfer between dinitrogenase reductase and the dinitrogenase. When this electron transfer occurs, the MgATP is hydrolyzed to yield MgADP plus orthophosphate. As ADP is inhibitory, one arranges experiments so that the ADP is promptly converted back to ATP so that ADP does not accumulate. This is accomplished conveniently by supplying creatine phosphate and creatine phosphokinase to reform the ATP.

After the dinitrogenase has accumulated adequate electrons (one at a time) it binds a substrate and effects its reduction. This process apparently is not the rate-limiting step in the nitrogenase system. The rate-limiting step occurs before this, and when it is clearly defined it will become an obvious point for a breeding attack to increase the rate of N_2 fixation. At this time one might think of increasing the concentration of dinitrogenase and dinitrogenase reductase and improving the production of ATP to increase overall N_2 fixation. An increased concentration of reductant to maintain the system in a highly reduced state also might improve fixation rates.

4. Reductants

The physiological reductants that most commonly transfer electrons to dinitrogenase reductase are ferredoxin and flavodoxin, with ferredoxin being more commonly functional. Ferredoxin has an unusually low oxidation-reduction potential; its potential is near that of the H_2 electrode. This potent reductant is capable of effecting the reduction of dinitrogenase reductase. *C. pasteurianum* when grown on a medium deficient in iron will produce flavodoxin, and it can function in place of ferredoxin.[15] Sometimes the rate of fixation in reconstructed systems can be increased by adding flavodoxin. However, in reconstructed systems one usually supplies sodium dithionite ($Na_2S_2O_4$) as the electron donor, as it is far more convenient to use than ferredoxin or flavodoxin. It can transfer electrons directly to dinitrogenase reductase in the absence of ferredoxin and flavodoxin. Although $Na_2S_2O_4$ has an extremely low potential, it is quite unable to transfer electrons directly to dinitrogenase. There is a specificity of electron transfer between dinitrogenase reductase and dinitrogenase, and hence the nitrogenase system requires the two proteins. The activity of a nitrogenase system can be decreased by depleting its ferredoxin, and the activity can be restored by adding back ferredoxin or a reductant such as $Na_2S_2O_4$. This suggests the possibility of improving the rate of N_2 fixation by enhancing the concentration of a reductant such as ferredoxin.

C. Substrates

1. N_2

The physiological substrate of major interest is N_2. N_2 is abundant in the atmosphere, but the bond in the molecule is extremely stable; hence there must be a highly energetic system to effect the splitting of the bond and the reduction of N_2 to ammonia. The Michaelis constant for N_2 varies among organisms between about 0.02 and 0.20 atm.[16] It is curious that cell-free preparations, which have no membrane barriers, exhibit a higher Michaelis constant than corresponding intact organisms. This has been observed with several N_2-fixing agents and suggests that somehow the intact organism has a way of concentrating the N_2 at the enzyme site to enhance the overall activity. The product of N_2 reduction is ammonia. The ammonia formed can be assimilated into organic compounds and utilized by the N_2-fixing agents.

The nitrogenase system, however, is not limited to the reduction of N_2 but reduces a considerable variety of substrates. These include protons (H^+), acetylene (C_2H_2), azide (N_3^-), nitrous oxide (N_2O), cyanide (CN^-), methyl isocyanide (CH_3NC), and cyclopropene $\begin{smallmatrix} H-C-H \\ (HC\!\!=\!\!\!=\!\!CH) \end{smallmatrix}$.

2. H^+

Protons always are present in the aqueous medium in which nitrogenase operates, and hence a reduction of H^+ to H_2 always is observed. In a normally functioning nitrogenase system there is approximately one H_2 produced for each N_2 reduced to $2\,NH_3$. The production of H_2 requires two electrons and the reduction of N_2 requires six electrons; so if equimolar amounts react, then 25% of the electrons will be utilized in the production of H_2. If the H_2 escapes from the system, this constitutes a 25% loss of the reducing energy invested. Considerable attention has been paid to the possibility of abolishing the production of H_2, but without success. In fact, kinetic evidence suggests that even an infinite concentration of N_2 will not completely block the production of H_2.[17]

As there is no known way to block H_2 production, one can consider the alternative approach of recapturing a part of the energy from H_2 through the activity of hydrogenase. Hydrogenase frequently is present in N_2-fixing systems, and hydrogenase has the potential to reduce a carrier such as ferredoxin or to regenerate MgATP which then can furnish energy to the nitrogenase system.[18] Investigation of the role of hydrogenase has been directed to the symbiotic system particularly in research with soybean plants.[19] Strains of *Rhizobium japonicum* possessing and lacking hydrogenase have been cultured, and it has been possible to demonstrate that the hydrogenase-containing strains will support a higher yield of soybeans or a more efficient use of the photosynthate than strains without hydrogenase.[19] Manipulation of hydrogenase has been particularly attractive among breeders trying to improve N_2 fixation. As the nitrogenase components reside in the bacteria, attention has been directed primarily to the bacteria rather than to the higher plant.

3. C_2H_2

The nitrogenase system is capable of reducing C_2H_2 to C_2H_4;[20,21] reduction requires two electrons. Although this process has little physiological significance, the reaction is important because it furnishes one of the simplest and most sensitive ways of measuring nitrogenase activity. Ethylene produced from acetylene can be measured quantitatively with great ease and speed by a gas chromatographic unit equipped with a flame ionization detector. Extremely small amounts of ethylene can be measured, and the system has the attractions of low cost, speed, simplicity, and high sensitivity. Used judiciously it can be a great help in investigations of N_2 fixation; the simple apparatus is particularly welcome in the field.

Interpreting acetylene reduction in terms of N_2 reduction involves certain precautions. It has been customary to divide the amount of the acetylene reduced by three to give the equivalent amount of N_2 reduced. This is based on the facile assumption that because the reduction of N_2 requires six electrons and the reduction of C_2H_2 requires two electrons there must be a direct 3-to-1 relationship. However, the situation is complicated by the fact that the N_2-fixation system in the presence of N_2 always utilizes electrons to generate H_2. In the presence of 10% acetylene in contrast to N_2, the production of H_2 is greatly suppressed; hence the assumption of a 3 to 1 ratio becomes invalid.[22] More commonly the ratio determined objectively with ^{15}N as a tracer is nearer to 4. In fact, the ratio is rather variable and must be determined under the specific conditions of the experiment to be valid.

4. Miscellaneous Substrates

Miscellaneous substrates such as azide, nitrous oxide, cyanide, methyl isocyanide, and cyclopropene have been studied less extensively. Investigations of each, however, have given hints as to the mechanism of the nitrogenase reaction. Azide, cyanide, and methyl isocyanide are noncompetitive inhibitors of N_2 reduction,[23] but are mutually competitive among themselves. Nitrous oxide is a competitive inhibitor of N_2.[23] These miscellaneous substrates warrant more thorough investigation, as they doubtless can yield new insight into the nature of the nitrogenase reaction.

D. Inhibitors

1. Competitive

Nitrous oxide and H_2 are the only two inhibitors that have shown clear-cut competitive inhibition of the nitrogenase system, but NO probably is competitive as well.[24] Nitrous oxide, in addition to being a competitive inhibitor, is a substrate for the nitrogenase system.

H_2 not only is competitive but also is specific in the sense that it has no influence on the utilization of fixed nitrogen compounds by N_2-fixing agents. The inhibition is strictly reversible. H_2 is of particular interest; not only is it a competitive inhibitor but it also is a product of the N_2-fixing system, and in the presence of hydrogenase it can function to support nitrogenase. The inhibitory constant K_i varies among the N_2-fixing agents. As H_2 is a competitive inhibitor, the inhibition obviously varies as the partial pressure of N_2 varies.

Metabolism of H_2 also generates interest, because the nitrogenase system in the presence of D_2 can generate HD.[25] Interest arises because HD formation is dependent upon the presence of N_2. It appears probable that HD is released from some labile intermediate in the N_2-fixation process. HD formation requires all of the same components and has the same limitations as N_2 fixation, so HD may well rise from an intermediate in N_2 fixation.

2. Noncompetitive

There are a number of noncompetitive inhibitors of N_2 fixation. These include the substrates that we have discussed such as acetylene, azide, cyanide, methyl isocyanide, and cyclopropene.[23,26] All of these substrates compete for electrons from the same electron pool of reduced dinitrogenase; hence, they are all mutually inhibitory.

Distinct from these competing substrates, CO is a notable inhibitor of the nitrogenase system. It is an extremely effective noncompetitive inhibitor, as a fraction of a percent of CO will virtually abolish N_2 fixation. It is particularly interesting that CO will block the reduction of all of the substrates for nitrogenase with the exception of H^+. Somehow H^+ escapes the CO block and continues to be reduced to H_2.[23]

3. Uncompetitive

The only uncompetitive inhibitor of nitrogen fixation that has been described is O_2. Oxygen irreversibly inactivates nitrogeanse. However, it is possible to set up a system with particles from *Azotobacter vinelandii* that will support nitrogenase activity while retaining respiratory protection against inactivation by O_2. Under these special circumstances one can add various concentrations of O_2 without inactivating the system and demonstrate that the reversible inhibition by O_2 is uncompetitive in nature.[27]

III. REGULATION OF NITROGEN FIXATION

Nitrogenase activity in the cell is regulated at the level of transcription, translation, substrate supply, and by covalent modification of dinitrogenase reductase. Of the modes of regulation listed above, only regulation of expression at the level of transcription has been observed in all cases.

A. Regulation of Transcription

In 1942, Wilson et al.[28] first unambiguously observed repression of nitrogenase activity in the free living organism *A. vinelandii*. Using ^{15}N as the tracer of nitrogen fixation it was shown that *A. vinelandii* would use fixed nitrogen compounds such as ammonia, nitrate, and urea in preference to N_2. In addition, an adaptation period was required before growth on N_2 would take place if the cells had been previously grown on fixed N. Although these experiments predated Monod's statement of the operon model, it was accepted from the time of these studies that nitrogen fixing organisms would use fixed N sources over molecular N_2. These observations were extended by Pengra and Wilson[29] and Strandberg and Wilson[30] who demonstrated that if ammonia was added to cultures of actively fixing *A. vinelandii*, nitrogenase activity as measured in crude extracts was lost over a period of 4 hr and growth continued at an increased rate.

Several pieces of information were obtained, and techniques were developed which were critical to understanding of the nitrogenase enzyme system and which allowed the regulation of the system to be studied. The first of these was the development of the acetylene reduction technique[31,32] which allowed the measurement of nitrogenase in vivo and thus allowed facile detection of the enzyme. Because of the extreme sensitivity of this assay it is known that nitrogenase is at least 100,000-fold repressed in the presence of fixed nitrogen. The realization that nitrogenase consists of two protein components one of which has two different subunit types[33,34] defined a minimum number of genes for the system. The finding that the phage Pl could operate as a generalized transducing phage in the N_2 fixing organism *Klebsiella pneumoniae* allowed mapping of the *nif* region.[35] In the last decade, increasingly sophisticated methods have been used to analyze the *nif* region of the *K. pneumoniae* genome. *K. pneumoniae* has become the organism of choice for studies of the genetics of *nif* because of the similarity of its genetic map to that of *Escherichia coli* and the ability to readily study the genome with methods and agents developed for *E. coli*. Information regarding *nif* genetics has

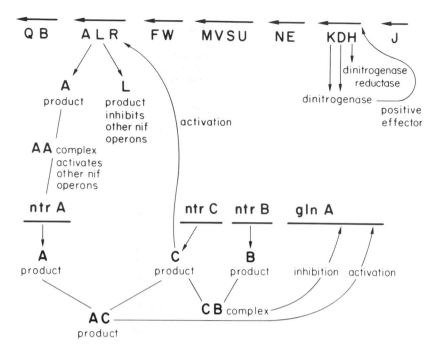

FIGURE 2.   Model for *regulation of expression* of *nif, gln, ntr* genes (adapted from Brill[36]).

also come from studies on *A. vinelandii* and more recently the rhizobia and the cyano-
bacterium *Anabaena cylindrica.*

The map of the *K. pneumoniae nif* region that has been developed is shown below
in Figure 2. The *nif* region of *K. pneumoniae* has been defined by selecting *nif*- mutants
and mapping them by transduction,[35] deletion mapping,[37-39] and transposon mapping.[40] The *nif* region is a 23-kilobase region consisting of 7 operons and 15 to 17 genes.
Not all of the operons are transcribed in the same direction. At least thirteen *nif*-
specific gene products have been identified at least as protein spots on two-dimensional
gels, and the functions of most of the gene products are known.[41]

The primary regulation of the expression of the *nif* region is by levels of fixed nitrogen and oxygen available to the cells. High levels of good nitrogen sources (ammonia,
glutamine, asparagine, and, in some cases, nitrate) result in repression of nitrogenase
activity while poor nitrogen sources (N_2, glutamate, and histidine) result in synthesis
of the *nif* gene products. High levels of dissolved oxygen also repress transcription of
the *nif* operons including the *nif* HDK operon which contains the coding information
for the oxygen-labile dinitrogenase- and dinitrogenase-reductase proteins. Various species have evolved a number of biochemical and physiological solutions to the oxygen
problem.[42-44]

Although the effect of fixed nitrogen on nitrogenase activity was observed very
early, understanding of the molecular mechanism of the effect is very recent. The glutamine synthetase (GS) inhibitor methionine sulfoximine was found to derepress nitrogenase in the presence of ammonia,[45] and one class of revertants of glutamine auxo-
trophs was found to express nitrogenase in the presence of ammonia.[46] In each case,
glutamine added to the medium repressed nitrogenase. These experiments implicated
GS in the regulation of *nif*. The regulation of GS itself is complex and involves (1)
feedback regulation of the enzyme by many small molecules, (2) regulation of the
enzyme by a cascade which results in adenylation of GS, and (3) regulation of the level
of expression of GS.[47] The present model for expression of the GS structural gene is

Table 1

FUNCTIONS AND PROPERTIES OF THE *NIF* GENE PRODUCTS

Dinitrogenase reductase	*nif* H gene product, mol wt 35,000
Dinitrogenase	*nif* D - subunit, mol wt 56,000
	nif K - subunit, mol wt 60,000
FeMo cofactor	*nif* N - mol wt 50,000
	nif E - mol wt 46,000
	nif B - unknown
	nif Q - unknown
Electron transport	*nif* J - pyruvate:flavodoxin oxidoreductase, mol wt 120,000
	nif F - flavodoxin
Regulation	*nif* A - activator of transcription, mol wt 60,000
	nif L - repressor or inhibitor of transcription, mol wt 50,000
	nif R - presumptive binding site for ntrC gene product; no gene product of *nif* R is known
Processing	*nif* M - processing of dinitrogenase reductase protein, mol wt 28,000
	nif V - processing of dinitrogenase protein, mol wt 42,000
	nif S - processing of dinitrogenase reductase protein
Unknown	*nif* W, *nif* U

that it is controlled by the ntrC, ntrB, and ntrA gene products.[48] The ntrC and ntrB genes are close to the structural gene for GS (glnA), while the ntrA gene is not. A region controlling the expression of all nitrogenase gene products was first detected by mutation[49,50] and found to map fairly distant from the structural genes for dinitrogenase and dinitrogenase reductase. This region, originally designated *nif* A, has been divided into three regions: *nif* A, *nif* L, and *nif* R. No gene product for *nif* R has been identified and it is presumed to be a promoter region. The ntrC and ntrA genes have been implicated in the regulation of *nif*.[51,52] The ntrC gene is thought to bind to the *nif* R region and allow transcription of the *nif* LA operon. The *nif* L and *nif* A gene products are thought to be repressor (or inhibitor) and an activator, respectively, of the other *nif* operons. The action of *nif* A also requires the presence of an active ntrA gene.[65,66] The *nif* A gene product and the ntrC gene product appear to be related proteins based on sequence homologies.[51,52] and in fact, the *nif* A gene product will substitute for the ntrC gene product in vivo if it is present at high concentrations. When 6-cyanopurine is added to solid media on which *K. pneumoniae* is plated, a purple pigment is generated in colonies which are expressing *nif*.[53] Analysis of this phenomena demonstrated that only the *nif* A gene must be expressed in order to observe the purpling; the expression of other *nif* genes is not required. The biochemistry of this observation has not been explained.

The model for *nif* expression is shown in Figure 2. It should be emphasized that this model rests on genetic rather than biochemical evidence and the mechanism by which the various gene products regulate expression of *gln* and *nif* are not known at this time.

The proposal that *nif* L gene product mediates oxygen repression comes from several sources. It was known that *K. pneumoniae* would fix N₂ only under anaerobic or microaerobic conditions.[54] Because of the oxygen lability of the enzymes involved, it was not clear that this was due to repression of the enzyme rather than destruction or inhibition of the enzyme once synthesized. The repression of enzyme synthesis by oxygen was demonstrated by showing that no enzyme activity or cross-reacting material was found when cells were grown under oxygen.[55,56] The *nif* L gene was defined by mutations, and some of these mutants lost the ability to respond to oxygen.[57,58] Because mutations in *nif* L often exhibit a polar effect on an expression on the positive effector *nif* A, many *nif* L mutants lose all ability to express nitrogenase. The *nif* L gene product has been identified on gels, but it has not been purified and its properties are not known.

There apparently are regulatory interactions among the *nif* operons in addition to the control by the *nif* LA region. It has been proposed that the *nif* HDK operon is autogenously regulated by the completed dinitrogenase protein. Some mutant strains that lack dinitrogenase show decreased synthesis of dinitrogenase reductase; this cannot be a polar effect as the H gene which codes for the dinitrogenase reductase precedes the K and D genes. Thus it is proposed that the dinitrogenase protein regulates HDK transcription. It is proposed that the completed dinitrogenase protein is required, as molybdenum starvation also results in decreased synthesis of the nitrogenase proteins.[59]

A potential for regulation also exists during the processing of the two nitrogenase proteins. The products of the *nif* M, V, S, and U genes have been proposed as potential processing steps of the other nitrogenase proteins. One class of mutants in *nif* V are reported to change the substrate specificity of the enzyme.[60]

Other, general regulators of gene expression may have some role in regulation of *nif* expression. Although there is no evidence of cAMP-mediated catabolite repression for the expression of *nif*, there is a correlation between ppGpp levels in the cell with nitrogen starvation and/or nitrogenase activity. ppGpp levels were observed to increase dramatically during nitrogenase induction in *K. pneumoniae*[61] and ppGpp levels were observed to drop rapidly upon addition of ammonia to nitrogen fixing cultures of *A. vinelandii, K. pneumoniae,* and *Clostridium pasteurianum.*[62] These effects may be mediated through expression of *ntr* genes or other unknown mechanisms. A pleiotropic effect of mutations in a number of *his* genes has also been observed on the expression of *nif* in *K. pneumoniae.*[63]

B. Regulation of Translation

Relatively little information exists about turnover of messenger RNA for *nif* gene products and *nif* proteins. Estimates of half-lives of nitrogenase proteins vary from several minutes in *K. pneumoniae,*[64,104] to many hours in *R. palustris.*[65] There does appear to be an effect of oxygen on the lifetime of the messenger RNA for the message for the HDK operon.[64,104]

C. Feedback Regulation of Nitrogenase

Ammonia is the first stable, fixed product of N_2 reduction by nitrogenase but there is no evidence of feedback inhibition of the enzyme by this compound. Zelitch[66] showed that cells of actively fixing cultures of *C. pasteurianum* would excrete $^{15}NH_4^+$. Most nitrogen-fixing organisms do not respond to the addition of ammonia by ceasing nitrogen-fixation activity as measured by the acetylene reduction assay (however, see sections below on the effects of ammonia on membrane potential in *A. vinelandii* and the modification of dinitrogenase reductase in photosynthetic bacteria). Ammonia feedback inhibition of nitrogenase in whole cells of *K. pneumoniae, C. pasteurainum,* and *A. vinelandii* was reexamined by Gordon et al.[67] In this study, the total ^{15}N fixed was examined in the intracellular and excreted fractions, and although added ammonia induced more ^{15}N excretion, it did not inhibit nitrogenase activity. No inhibitory effect of ammonia on isolated nitrogenase has been reported.

Because of its central role in pyrimidine biosynthesis, carbamyl phosphate has been investigated as a potential regulator of nitrogenase. Seto and Mortenson[68] observed that carbamyl phosphate inhibited nitrogenase in vitro with a maximal inhibition of approximately 50%. Gordon et al.[67] investigated carbamyl phosphate as an inhibitor in crude extracts and they observed little inhibition even at 10 mM carbamyl phosphate.

MgATP is required for nitrogenase activity and is hydrolyzed to MgADP and P_i.[69,70] MgADP is a potent inhibitor of nitrogenase activity[71] and thus may play a role in regulation of nitrogenase activity in vivo. The K_m for ATP is reported to be in the range of 80 to 200 μM in various species and the K_i for MgADP is 20 μM (see Section II.B.3 of this review). The paradox of regulation by MgADP is that inhibitory effects of MgADP can be demonstrated in vitro but when measurements of nucleotide pools in whole cells are made, clearly inhibitory levels of ADP are observed with no apparent inhibition of nitrogenase activity in vivo. Based on measurements made in vitro, nitrogenase activity would be expected to be more sensitive to ATP/ADP ratio and energy charge than most other enzymes, but very little correlation of nitrogenase activity with energy charge is observed. Davis and Kotake[72] have calculated that the ADP/ATP ratios observed in a variety of organisms should give a 90% or greater inhibition. They further estimate that nitrogenase is operating at near maximal rate in vivo despite the apparent unfavorable ADP/ATP ratio. Laane et al.[73] also have noted the lack of correlation of nitrogenase activity in vivo with energy charge and have proposed that membrane energization is the controlling factor for nitrogenase activity in vivo. They propose that membrane energization is the driving force for electron flow to nitrogenase in vivo and have demonstrated a correlation between membrane energization and nitrogenase activity. Davis and Kotake[72] make note of the possibility that membrane energization controls the free Mg^{+2} level on the cell, since ATP binds Mg^{+2} more tightly than ADP does. If Mg^{+2} were limiting, the effective ratio of MgATP/MgADP operating in the cell would be much higher than the pool measurements indicate. This proposal certainly deserves more study.

Although direct feedback inhibition of nitrogenase activity by ammonia may not occur, some transitory effects of ammonia on whole cell activity have been observed.[74] These effects have been attributed to the disruption of the membrane potential and resulting loss of electron flow to nitrogenase. The effect of ammonia has been reported in *A. vinelandii*, *Rhizobium* bacteroids, cyanobacteria, and the photosynthetic bacterium *Rhodopseudomonas sphaeroides*. In each case, the effect of ammonia on membrane potential has been demonstrated and correlated with loss of in vivo nitrogenase activity.

D. Regulation of Nitrogenase by Covalent Modification in the Purple Nonsulfur Photosynthetic Bacteria

Several nonsulfur photosynthetic bacteria exhibit regulation of nitrogenase activity by inactivation of dinitrogenase reductase. Kamen and Gest first noted that ammonia rapidly inhibited hydrogen evolution (now known to be nitrogenase dependent) and N_2 uptake by whole cells of *Rhodospirillum rubrum*.[75] When techniques for preparation of cell-free extracts became available, they were applied to *R. rubrum* and other photosynthetic bacteria and it was found that an unusually high Mg^{+2} concentration was required for activity[76] and that membrane fragments of the disrupted cells inhibited crude extract activity.[76] Several attempts to purify nitrogenase from *R. rubrum* were unsuccessful and it was observed that activity in crude extracts was low and nonlinear.[77] These observations were explained when it was found that the dinitrogenase reductase isolated from cells grown with N_2 or glutamate as the N source was inactive and required activation by an activating enzyme.[78] This reaction requires MgATP, a free divalent metal (Mn^{+2} is best), and the activating enzyme. The O_2-labile activating enzyme is found in the membrane fraction of the crude extract. Membranes which have been exposed to air inhibit the activation reactions.[79] Dinitrogenase reductase can be isolated in an active form from cells grown with limiting ammonia.[80] The dinitrogenase reductase from *R. rubrum* is similar to those from other organisms with respect to molecular weight, amino acid composition, iron and acid labile sulfur content, ability to bind MgATP, and EPR signal.[81] The nitrogenase proteins from *R. rubrum* also cross-react with nitrogenase components from other organisms.[82]

The inactive form of the dinitrogenase reductase from *R. rubrum* has attached to it a covalently bound modifying group which consists of phosphate, pentose, and adenine.[81] There is one modifying group per protein dimer. The modifying group is not AMP. Unlike the AMP found attached to glutamine synthase, the modifying group from *R. rubrum* dinitrogenase reductase is not removed by venom diesterase; the linkages of the components of the modifying group to each other and to the protein are not precisely known.

Dinitrogenase reductase can be activated in vitro in the presence of MgATP, free Mn^{2+}, or Mg^{2+}, and activating enzyme. Alternatively, if the enzyme is oxidized, MgATP is not required.[83] It is also possible to activate the protein using an obviously nonphysiological set of conditions; i.e., heating the protein to 50°C or higher.[84] The modifying group is removed during activating enzyme-dependent activation. The structure of the modifying group is not completely known and there may be components to it that have not been detected yet.

The modifying group is attached to one of two identical subunits of dinitrogenase reductase, and the two subunits can be separated on SDS acrylamide gels based on the presence of the modifying group on one of the subunits.[86] Activation by activating enzyme or by heating results in conversion of the slower moving, modified subunit to a faster moving subunit.

Although inactivation has not been detected in vitro, extracts can be made and analyzed for upper subunit composition of the protein. If the cells are grown in the presence of radioactively labeled precursors of the modifying group (e.g., ^{32}P), incorporation of label into the upper band can be taken as evidence of incorporation of the inactivation unit on the protein. If the extracts are made rapidly, the ratio of upper and lower subunit will reflect the amount of active and inactive protein in vivo.[87] This approach has shown that not only ammonia, but also darkness, uncouplers such as CCCP, and redox dyes such as methylene blue result in modification of dinitrogenase reductase in vivo.[87] The incorporation of ^{14}C label from ATP and ^3H label from adenine into the protein by whole cells has been demonstrated.[88,89] However, ^{32}P label from the α position of ATP did not appear in dinitrogenase reductase, again indicating that AMP is not the modifying group.

IV. SYMBIOTIC N_2 FIXATION IN LEGUMES

A. Plant-Bacterial Interactions

In the symbiotic N_2-fixation system in leguminous plants there is a complex interaction between the bacteria and the plant. One must recognize that this makes a breeding approach to improving N_2 fixation a rather complex operation. It has been established that the genetic information for the N_2 fixation system resides in the bacteria. This point was not clear for many years, but it now has been possible under proper conditions to induce isolated rhizobia to fix N_2 apart from the host plant.[90-92] The conditions for expression of the nitrogenase are rather specific, but there now is no difficulty in showing that the complete genetic information for nitrogenase is carried by the bacteria.

Special conditions[90,92] necessary for expression of nitrogenase in free-living rhizobia are, first, a proper substrate (this is supplied by the plant in the symbiotic system; organic acids can function). The second requirement is a nitrogen source (free-living rhizobia require some N in addition to N_2) which does not suppress the nitrogenase system; glutamate will serve. Third, a proper concentration of O_2 is required. Isolated rhizobia will fix N_2 only at a rather low pO_2, but an adequate supply of O_2 is furnished by the leghemoglobin. The leghemoglobin facilitates diffusion of O_2 into the nodule but releases it at such a low concentration that it will not inactive the nitrogenase system but will be adequate to support oxidative phosphorylation for generation of ATP.[93]

It generally is conceded that ammonia produced in the plant is added to glutamate to form glutamine as catalyzed by glutamine synthetase.[94] Glutamine in turn can interact with α-keto glutarate in the presence of glutamate synthase to form two molecules of glutamate. The glutamate then can serve as a donor in a wide variety of transamination reactions to form other amino acids. Certain plants transfer their nitrogen within the plant primarily in the form of amides and others transfer the nitrogen primarily as ureides.[95]

As mentioned, the interactions between the bacteria and the plant may be rather complex. It is long established that individual nodulating strains on a specific plant vary widely in their effectiveness in N_2 fixation. Likewise a single strain of bacteria on a variety of closely related plants will show a marked difference in ability to fix N_2. These observations gave rise to the concept of strain variation and host-plant specificity among plants and bacteria in the symbiotic N_2 fixation association.[96] These complexities pose problems in any program aimed at breeding for a more effective N_2 fixation association.

B. Energy from Photosynthesis

One of the great advantages of the symbiotic N_2-fixing system is that it can be driven by the energy from photosynthesis and involves no expenditure of energy from fossil fuels. Photosynthesis generates ATP in abundance via its photochemical reactions,[97] and it is possible that some of this ATP can be used directly to drive the N_2 fixation system. Alternatively, photosynthate produced by the photosynthesizing plant can be metabolized by the bacteria to generate the ATP and reductant necessary to drive N_2 fixation. Considerable current research centers around attempts to improve the efficiency of photosynthesis, as plants vary widely in their efficiency.[98] The most notable difference is between the C_4 and the C_3 plants. To date, it has not been possible to transfer C_4 characteristics to a C_3 plant to improve its photosynthetic effectiveness. Such an alteration could be adapted to the legumes when successful, as all legumes are C_3 plants.

C. Enhancement of N_2 Fixation
1. Plant

The role of the plant in symbiotic N_2 fixation has been neglected relative to the bacteria. The bacteria carry the genetic information for the nitrogenase system, but they cannot function properly without the support of the plant. The plant, through photosynthesis, furnishes energy for driving the N_2 fixation process. The bacteria are lodged in a root nodule where the partial pressure of O_2 is effectively controlled and where they receive photosynthate directly through the vascular system of the plant.[99] In turn, the fixed nitrogen can be exported directly through the vascular system to the rest of the plant. This arrangement permits the development of a very high concentration of bacteria in the nodule, and a high concentration means that substantial N_2 fixation can be supported. This confers a great advantage over "associative N_2 fixation" in which the growth of the bacteria and fixation are dependent upon photosynthate that leaks from the root system, and in which the fixed nitrogen must enter the plant by some route not directly connected to the vascular system.[100] Studies of host-plant specificity show how important the proper plant is in achieving effective N_2 fixation. Obviously the plant can be a prime target for breeding improved N_2 fixation associations.

2. Bacteria

The bacteria carry the genetic information for N_2 fixation. Hence, if one can improve their fixation it will be beneficial to the plant-bacterial association. Once again the process is not simple, because the nitrogenase system is a complex of proteins, an ATP-generating system, and perhaps a H_2 recycling system. Any one of these may be a target for genetic manipulation, but the components of the whole system must be maintained in harmony to keep it highly effective. One can study fixation by rhizobia apart from the host plant, but improving their system under these rather artificial conditions is no guarantee that they will function more effectively in the plant-bacterial association.

D. Regulation in the Symbiotic System

There are many possible points of regulation in symbiotic systems such as the *Rhizobium:* legume symbiosis. These include the regulation of events in establishing the symbiosis as well as regulation of the active symbiosis. It is clear that fixed nitrogen has an effect on the *Rhizobium:* legume symbiosis at at least two points: (1) added ammonia or nitrate will inhibit nodule formation, and (2) added ammonia or nitrate will inhibit the activity of nodules that are present and active on the root systems.[101] The biochemical basis for these effects is not completely understood and genetic analysis of the events during nodule formation are just being explored.

Dazzo et al. have demonstrated that a specific lectin, trifoliin, is not as abundant on the roots of clover plants grown in the presence of nitrate or ammonia.[102] This lectin is thought to be invovled in the initial recognition between the proper symbiont and the plant. The inhibition of nodule activity by fixed nitrogen often has been observed. One hypothesis that has been advanced to explain inhibition of nodule activity by nitrate is that nitrate is reduced to nitrite by nitrate reductase and that the accumulated nitrite is toxic to the nodule. However, this does not explain inhibition of nodule activity by ammonia.

E. Progress to Date

Plant breeders have been interested in the possibility of improving symbiotic N_2 fixation in leguminous plants.[103] There has been relatively little attention to the nonlegumes such as the alder, *Ceanothus* sp., *Myrica* sp., etc. The work generally has centered on the bacteria, because they carry the genetic information for N_2 fixation, but one cannot neglect the plant, because it influences the symbiotic relationship markedly. Mutants have been made of N_2-fixing rhizobia that have given some enhanced early growth of legumes. These strains, however, have not supported the production of a larger crop of legume seeds or of forage. As the crop yield is of primary interest to the farmer, improving the growth of the plants to the 30-day stage has had little impact on practical agriculture. However, the genetic manipulations accomplished to date have led the way, and improvements that will be accompanied by an increased yield of crops can be anticipated.

An area that has been particularly promising has been the manipulation of hydrogenase in the symbiotic N_2-fixing system. The loss of H_2 from nitrogenase can dissipate a substantial amount of energy. If this energy can be recaptured in part by recycling H_2 to form ATP or a strong reductant, it should be possible to enhance crop yields. Although this approach has not been uniformly successful in the field, the work in Evans' laboratory[19] has demonstrated that crop yields can be improved by strains of hydrogenase positive root nodule bacteria in comparison with hydrogenase negative strains.

Research in the area of genetic manipulation to improve biological N_2 fixation is attractive, and the system presents a number of points for logical attack. As the research is expanded, one can anticipate that success will follow, so that the system which has been evolved over the millenia will be altered rather rapidly to improve its effectiveness.

REFERENCES

1. Paul, E. A., Contribution of nitrogen fixation to ecosystem functioning and nitrogen fluxes on a global basis, in *Environmental Role of Nitrogen-Fixing Blue-Green Algae and Asymbiotic Bacteria,* Granhall, U., Ed., *Ecological Bulletin,* 26, 1978, 282.
2. Burris, R. H., Arp, D. J., Hageman, R. V., Houchins, J. P., Sweet, W. J., and Tso, M.-Y., Mechanism of nitrogenase action, *Proc. 4th Int. Symp. Curr. Perspec., Nitrogen Fixation,* Gibson, A. H. and Newton, W. E., Eds., Elsevier, Amsterdam, 1980, 56.
3. Eady, R. R., Methods for studying nitrogenase, in *Methods for Evaluating Biological Nitrogen Fixation,* Bergersen, F. J., Ed., John Wiley & Sons, New York, 1980, 213.
4. Orme-Johnson, W. H., Hamilton, W. D., Ljones, T., Tso, M.-Y., Burris, R. H., Shah, V. K., and Brill, W. J., Electron paramagnetic resonance of nitrogenase and nitrogenase components from *Clostridium pasteurianum W5 and Azotobacter vinelandii* OP, *Proc. Natl. Acad. Sci. U.S.A.,* 69, 3142, 1972.
5. Shah, V. K. and Brill, W. J., Isolation of an iron-molybdenum cofactor from nitrogenase, *Proc. Natl. Acad. Sci. U.S.A.,* 74, 3249, 1977.
6. Bui, P. T. and Mortenson, L. E., Mechanism of the enzymic reduction of N_2: the binding of adenosine 5'-triphosphate and cyanide to the N_2-reducing system, *Proc. Natl. Acad. Sci. U.S.A.,* 61, 1021, 1968.
7. Tso, M.-Y. W. and Burris, R. H., The binding of ATP and ADP by nitrogenase components from *Clostridium pasteurianum, Biochim. Biophys. Acta,* 309, 263, 1973.
8. Zumft, W. G., Mortenson, L. E., and Palmer, G., Electron paramagnetic resonance studies on nitrogenase. Investigation of the oxidation-reduction behavior of azoferredoxin and molybdoferredoxin with potentiometric and rapid-freeze techniques, *Eu. J. Biochem.,* 46, 525, 1974.
9. Thorneley, R. N. F., Yates, M. G., and Lowe, D. J., Nitrogenase of *Azotobacter chroococcum.* Kinetics of the reduction of oxidized iron-protein by sodium dithionite, *Biochem. J.,* 155, 137, 1976.
10. Ljones T. and Burris, R. H., Evidence for one-electron transfer by the iron protein of nitrogenase, *Biochem. Biophys. Res. Commun.,* 80, 22, 1978.
11. Winter, H. C. and Burris, R. H., Stoichiometry of the adenosine triphosphate requirement for N_2 fixation and H_2 evolution by a partially purified preparation of *Clostridium pasteurianum, J. Biol. Chem.,* 243, 940, 1968.
12. Daesch, G. and Mortenson, L. E., Sucrose catabolism in *Clostridium pasteurianum* and its relation to N_2 fixation, *J. Bacteriol.,* 96, 346, 1968.
13. Hageman, R. V. and Burris, R. H., Nitrogenase and nitrogenase reductase associate and dissociate with each catalytic cycle, *Proc. Natl. Acad. Sci. U.S.A.,* 75, 2699, 1978.
14. Emerich, D. W. and Burris, R. H., Interactions of heterologous nitrogenase components that generate catalytically inactive complexes, *Proc. Natl. Acad. Sci. U.S.A.,* 73, 4369, 1976.
15. Knight, E., Jr. and Hardy, R. W. F., Isolation and characteristics of flavodoxin from nitrogen-fixing *Clostridium pasteurianum, J. Biol. Chem.,* 241, 2752, 1966.
16. Hwang, J. C. and Burris, R. H., Nitrogenase-catalyzed reactions, *Biochim. Biophys. Acta,* 283, 339, 1972.
17. Hwang, J. C., Chen, C. H., and Burris, R. H., Inhibition of nitrogenase-catalyzed reductions, *Biochim. Biophys. Acta.* 292, 256, 1973.
18. Peterson, R. B. and Burris, R. H., Hydrogen metabolism in isolated heterocysts of *Anabaena* 7120, *Arch. Microbiol.,* 116, 125, 1978.
19. Evans, H. J., Emerich, D. W., Ruiz-Argüeso, T., Maier, R. J., and Albrecht, S. L., Hydrogen metabolism in the legume- *Rhizobium* symbiosis, in *Nitrogen Fixation,* Vol. 2, Newton, W. E. and Orme-Johnson, W. H., Eds., University Park Press, Baltimore, 1980, 325.
20. Dilworth, M. J., Acetylene reduction by nitrogen-fixing preparations from *Clostridium pasteurianum, Biochim. Biophys. Acta,* 127, 285, 1966.
21. Schöllhorn, R. and Burris, R. H., Study of intermediates in nitrogen fixation, *Fed. Proc. Fed. Am. Soc. Exp. Biol.,* 25, 710, 1966.

22. Burris, R. H., The acetylene-reduction technique, in *Nitrogen Fixation by Free-Living Micro-organisms*, Stewart, W. D. P., Ed., International Biological Program, Vol. 6, Cambridge University Press, Cambridge, 1975, 249.

23. Rivera-Ortiz, J. M. and Burris, R. H., Interactions among substrates and inhibitors of nitrogenase, *J. Bacteriol.*, 123, 537, 1975.

24. Lockshin, A. and Burris, R. H., Inhibitors of nitrogen fixation in extracts from *Clostridium pasteurianum*, *Biochim. Biophys. Acta*, 111, 1, 1965.

25. Hoch, G. E., Schenider, K. C., and Burris, R. H., Hydrogen evolution and exchange, and conversion of N_2O to N_2 by soybean root nodules, *Biochim. Biophys. Acta*, 37, 273, 1960.

26. McKenna, C. E. and Huang, C. W., *In vivo* reduction of cyclopropene by *Azotobacter vinelandii* nitrogenase, *Nature (London)*, 280, 609, 1979.

27. Wong, P. P. and Burris, R. H., Nature of oxygen inhibition of nitrogenase from *Azotobacter vinelandii*, *Proc. Natl. Acad. Sci. U.S.A.*, 69, 672, 1972.

28. Wilson, P. W., Hull, J. F., and Burris, R. H., Competition between free and combined nitrogen in nutrition of *Azotobacter*, *Proc. Natl. Acad. Sci. U.S.A.*, 29, 289, 1943.

29. Pengra, R. M. and Wilson, P. W., Physiology of nitrogen fixation by *Aerobacter aerogenes*, *J. Bacteriol.*, 75, 21, 1958.

30. Strandberg, G. W. and Wilson, P. W., Formation of the nitrogen-fixing enzyme system in *Azotobacter vinelandii*, *Can. J. Microbiol.*, 14, 25, 1968.

31. Stewart, W. D. P., Fitzgerald, G. P., and Burris, R. H., In situ studies on N_2 fixation using the acetylene reduction technique, *Proc. Natl. Acad. Sci. U.S.A.*, 58, 2071, 1967.

32. Koch, B. and Evans, H. J., Reduction of acetylene to ethylene by soybean root nodules, *Plant Physiol.*, 41, 1748, 1966.

33. Tso, M.-Y., Ljones, T., and Burris, R. H., Purification of the nitrogenase proteins from *Clostridium pasteurianum*, *Biochim. Biophys. Acta*, 267, 600, 1972.

34. Shah, V. K. and Brill, W. J., Nitrogenase IV. Simple method of purification to homogeneity of nitrogenase components from *Azotobacter vinelandii*, *Biochim. Biophys. Acta*, 305, 445, 1973.

35. Streicher, S., Gurney, E., and Valentine, R. C., Transduction of the nitrogen fixation genes in *Klebsiella pneumoniae*, *Proc. Natl. Acad. Sci. U.S.A.*, 68, 1174, 1971.

36. Brill, W. J., Biochemical genetics of nitrogen fixation, *Microbiol. Rev.*, 44, 449, 1980.

37. MacNeil, T., MacNeil, D., Roberts, G. P., Supiano, M. A., and Brill, W. J., Fine-structure mapping and complementation analysis of *nif* (nitrogen fixation) genes in *Klebsiella pneumoniae*, *J. Bacteriol.*, 136, 253, 1978.

38. Dixon, R., Eady, R. R., Espin, G., Hill, S., Iaccarino, M., Kahn, D., and Merrick, M., Analysis of regulation of *Klebsiella pneumoniae* nitrogen fixation *(nif)* gene cluster with gene fusions, *Nature (London)*, 286, 128, 1980.

39. Shanmugam, K. T., Loo, A. S., and Valentine, R. C., Deletion mutants of nitrogen fixation in *Klebsiella pneumoniae:* mapping of a cluster of *nif* genes essential for nitrogenase activity, *Biochim. Biophys. Acta*, 338, 545, 1974.

40. Espin, G., Alvarez-Morales, A., and Merrick, M., Complementation analysis of glnA-linked mutations which affect nitrogen fixation in *Klebsiella pneumoniae*, *Mol. Gen. Genet.*, 184, 213, 1981.

41. Roberts, G. P., MacNeil, T., MacNeil, D., and Brill, W. J., Regulation and characterization of protein products coded by the *nif* (nitrogen fixation) genes of *Klebsiella pneumoniae*, *J. Bacteriol.*, 136, 267, 1978.

42. Bergerson, F. J., Turner, G. L., and Appleby, C. A., Studies of the physiological role of leghaemoglobin in soybean root nodules, *Biochim. Biophys. Acta*, 292, 271, 1973.

43. Robson, R. L., Characterization of an oxygen stable nitrogenase complex isolated from *Azotobacter vinelandii*, *Biochem. J.*, 181, 569, 1979.

44. Haselkorn, R., Heterocysts, *Annu. Rev. Plant Physiol.*, 29, 319, 1978.

45. Brill, W. J., *nif*-Mutants of free-living bacteria, in *Proc. 1st Symp. Nitrogen Fixation*, Newton, W. E. and Nyman, C. J., Eds., Washington State University Press, Pullman, Washington, 1976, 327.

46. Gordon, J. K. and Brill, W. J., Mutants that produce nitrogenase in the presence of ammonia, *Proc. Natl. Acad. Sci. U.S.A.*, 69, 3501, 1972.

47. Tyler, B., Regulation of the assimilation of nitrogen compounds, *Annu. Rev. Biochem.*, 47, 1127, 1978.

48. Pahel, G., Rothstein, D. M., and Magasanik, B., Complex glnA-glnL-glnG operon of *Escherichia coli*, *J. Bacteriol.*, 150, 202, 1982.

49. St. John, R. T., Johnston, M., Seidman, C., Garfinkel, D., Gordon, J. K., Shah, V. K., and Brill, W. J., Biochemistry and genetics of *Klebsiella pneumoniae* mutant strains unable to fix N_2, *J. Bacteriol.*, 121, 759, 1975.

50. Dixon, R., Kennedy, C., Kondorosi, A., Krishnapillai, V., and Merrick, M., Complementation analysis of *Klebsiella pneumoniae* mutants defective in nitrogen fixation, *Mol. Gen. Genet.*, 157, 189, 1977.

51. Ow, D. W. and Ausubel, F. M., Regulation of nitrogen metabolism genes by *nif* A gene product in *Klebsiella pneumoniae, Nature (London),* 301, 307, 1983.
52. Drummond, M., Clements, J., Merrick, M., and Dixon, R., Positive control and autogenous regulation of the *nif* LA promoter in *Klebisella pneumoniae, Nature (London),* 301, 302, 1983.
53. MacNeil, D. and Brill, W. J., 6-Cyanopurine, a color indicator useful for isolating mutations in the *nif* (nitrogen fixation) genes of *Klebsiella pneumoniae, J. Bacteriol.,* 136, 247, 1978.
54. Klucas R., Nitrogen fixation by *Klebisella* grown in the presence of oxgyen, *Can. J. Microbiol.,* 18, 1845, 1972.
55. St. John, R. T., Shah, V. K., and Brill, W. J., Regulation of nitrogenase synthesis by oxygen in *Klebsiella pneumoniae, J. Bacteriol.,* 119, 266, 1974.
56. Eady, R. R., Issack, R., Kennedy, C., Postgate, J. R., and Ratcliffe, H. D., Nitrogenase synthesis in *Klebsiella pneumoniae:* comparison of ammonium and oxygen regulation, *J. Gen. Microbiol.,* 104, 277, 1978.
57. Kennedy, C., Cannon, F., Cannon, M., Dixon, R., Hill, S., Jensen, J., Kumar, S., McLean, P., Merrick, M., Robson, R., and Postgate, J., Recent advances in the genetics and regulation of nitrogen fixation, in *Current Perspectives in Nitrogen Fixation,* Gibson, A. H. and Newton, W. E., Eds., Australian Academy of Sciences, Canberra, 1981, 146.
58. MacNeil, D., Zhu, J., and Brill, W. J., Regulation of nitrogen fixation in *Klebsiella pneumoniae:* isolation and characterization of strains with *nif*-lac fusions, *J. Bacteriol.,* 145, 348, 1981.
59. Kahn, D., Hawkins, M., and Eady, R. R., Nitrogen fixation in *Klebsiella pneumoniae:* nitrogenase levels and the effect of added molybdate on nitrogenase derepressed under molybdenum deprivation, *J. Gen. Microbiol.,* 128, 779, 1982.
60. McLean, P. A. and Dixon, R. A., Requirement of *nif* V gene for production of wild-type nitrogenase enzyme in *Klebsiella pneumoniae, Nature (London),* 292, 655, 1981.
61. Riesenberg, D., Erdei, S., Kondorosi, E., and Kari, C., Positive involvement of ppGpp in derepression of the *nif* operon in *Klebsiella pneumoniae, Mol. Gen. Genet.,* 185, 198, 1982.
62. Kleiner, D. and Phillips, S., Relative levels of guanosine 5'-diphosphate 3'-diphosphate (ppGpp) in some N_2 fixing bacteria during derepression and repression of nitrogenase, *Arch. Microbiol.,* 128, 341, 1981.
63. Jensen, J. S. and Kennedy, C., Pleiotropic effect of his gene mutations on nitrogen fixation in *Klebsiella pneumoniae, EMBO J.,* 1, 197, 1982.
64. Kaluza, K. and Hennecke, H., Regulation of nitrogenase messenger RNA synthesis and stability in *Klebsiella pneumoniae, Arch. Microbiol.,* 130, 38, 1981.
65. Arp, D. and Zumft, W. G., Overproduction of nitrogenase by nitrogen-limited cultures of *Rhodopseudomonas palustris, J. Bacteriol.,* 153, 1322, 1983.
66. Zelitch, I., Simultaneous use of molecular nitrogen and ammonia by *Clostridium pasteurianum, Proc. Natl. Acad. Sci. U.S.A.,* 37, 559, 1951.
67. Gordon, J. K., Shah, V. K., and Brill, W. J., Feedback inhibition of nitrogenase, *J. Bacteriol.,* 148, 884, 1981.
68. Seto, B. and Mortenson, L. E., Mechanism of carbamyl phosphate inhibition of nitrogenase of *Clostridium pasteurianum, J. Bacteriol.,* 117, 805, 1974.
69. McNary, J. E. and Burris, R. H., Energy requirements for nitrogen fixation by cell-free preparations from *Clostridium pasteurianum, J. Bacteriol.,* 84, 598, 1962.
70. Mortenson, L. E., Ferredoxin and ATP requirements for nitrogen fixation in cell-free extracts of *Clostridium pasteurianum, Proc. Natl. Acad. Sci. U.S.A.,* 52, 272, 1964.
71. Tso, M-Y. W. and Burris, R. H., The binding of ATP and ADP by nitrogenase components from *Clostridium pasteurianum, Biochim. Biophys. Acta,* 309, 263, 1973.
72. Davis, L. C. and Kotake, S., Regulation of nitrogenase activity in aerobes by Mg_{+2} availability: an hypothesis, *Biochem. Biophys. Res. Commun.,* 93, 934, 1980.
73. Laane, C., Haaker, H., and Veeger, C., On the efficiency of oxidative phosphorylation in membrane vesicles of *Azotobacter vinelandii* and of *Rhizobium leguminosarum* bacteroids, *Eur. J. Biochem.,* 97, 369, 1979.
74. Haaker, H., Laane, C., Hellingwerf, K., Houwer, B., Konings, W., and Veeger, C., Short-term regulation of the nitrogenase activity in *Rhodopseudomonas sphaeroides, Eur. J. Biochem.,* 127, 639, 1982.
75. Kamen, M. D. and Gest, H., Evidence for a nitrogenase system in the photosynthetic bacterium *Rhodospirillum rubrum, Science,* 109, 560, 1949.
76. Burns, R. C. and Bulen, W. A., A procedure for the preparation of extracts from *Rhodospirillum rubrum* catalyzing N_2 reduction and ATP dependent H_2 evolution, *Arch. Biochem. Biophys.,* 113, 461, 1966.
77. Munson, T. O. and Burris, R. H., Nitrogen fixation by *Rhodospirillum rubrum* grown in nitrogen-limited continuous culture, *J. Bacteriol.,* 97, 1093, 1969.

78. Ludden, P. W. and Burris, R. H., Activating factor for the iron protein of nitrogenase from *Rhodospirillum rubrum, Science,* 194, 424, 1976.
79. Triplett, E. W., Wall, J. D., and Ludden, P. W., Expression of the activating enzyme and Fe protein of nitrogenase from *Rhodospirillum rubrum, J. Bacteriol.,* 152, 786, 1982.
80. Carithers, R. P., Yoch, D. C., and Arnon, D. I., Two forms of nitrogenase from the photosynthetic bacterium *Rhodospirillum rubrum, J. Bacteriol.,* 137, 779, 1979.
81. Ludden, P. W. and Burris, R. H., Purification and properties of nitrogenase from *Rhorospirillum rubrum* and evidence for phosphate, ribose and an adenine-like unit covalently bound to the iron protein, *Biochem. J.,* 175, 251, 1978.
82. Emerich, D. W. and Burris, R. H., Interactions of heterologous nitrogenase components that generate catalytically inactive complexes, *Proc. Natl. Acad. Sci. U.S.A.,* 73, 4369, 1976.
83. Ludden, P. W. and Burris, R. H., Removal of an adenine like molecule during activation of dinitrogenase reductase from *Rhodospirillum rubrum, Proc. Natl. Acad. Sci. U.S.A.,* 76, 6201, 1979.
84. Dowling, T. E., Preston, G. G., and Ludden, P. W., Heat activation of the Fe protein of nitrogenase of *Rhodospirillum rubrum, J. Biol. Chem.,* 257, 13987, 1982.
85. Gotto, J. W. and Yoch, D. C., Regulation of *Rhodospirillum rubrum* nitrogenase activity. Properties and interconversion of active and inactive Fe protein, *J. Biol. Chem.,* 257, 2868, 1982.
86. Preston, G. G. and Ludden, P W., Change in subunit composition of the Fe protein of nitrogenase from *Rhodospirillum rubrum* during activation and inactivation of the iron protein, *Biochem. J.,* 205, 489, 1982.
87. Ludden, P. W., Nordlund, S., Dowling, T. E., and Triplett, E., Regulation of nitrogenase activity by covalent modification, in *Current Topics in Plant Biochemistry and Physiology, 1983,* Proc. inaugural plant biochemistry and physiology symp., University of Missouri-Columbia, April 7 to 9, 1982, Randall, D. D., Blevins, D. G., and Larson, R., Eds., 1983, 204.
88. Vignais, P. M., Colbeau, A., Jouanneau, Y., Willison, J. C., and Michalski, W. P., H_2 metabolism in photosynthetic bacteria, Abstr. 4th Int. Symp. on Photosynthetic Procaryotes, Abstr. B61, 1982.
89. Nordlund, S. and Ludden, P. W., Incorporation of adenine into the modifying group of inactive iron protein of nitrogenase from *Rhodospirillum rubrum, Biochem. J.,* 209, 881, 1983.
90. Pagan, J. D., Child, J. J., Scowcroft, W. R., and Gibson, A. H., Nitrogen fixation by *Rhizobium* cultured on a defined medium, *Nature (London),* 256, 406, 1975.
91. Kurz, W. G. W. and LaRue, T. A., Nitrogenase activity in rhizobia in absence of plant host, *Nature (London),* 256, 407, 1975.
92. McComb, J. A., Elliott, J., and Dilworth, M. J., Acetylene reduction by *Rhizobium* in pure culture, *Nature (London),* 256, 409, 1975.
93. Bergersen, F. J., Leghaemoglobin, oxygen supply and nitrogen fixation: studies with soybean nodules, *Annu. Proc. Phytochem. Soc. Eur.,* 18, 139, 1980.
94. Miflin, B. J. and Lea, P. J., Amino acid metabolism, *Annu. Rev. Plant Physiol.,* 28, 299, 1977.
95. Pate, J. S., Atkins, C. A., White, S. T., Rainbird, R. M., and Woo, K. C., Nitrogen nutrition and xylem transport of nitrogen in ureide-producing grain legumes, *Plant Physiol.,* 65, 961, 1980.
96. Wilson, P. W., *The Biochemistry of Symbiotic Nitrogen Fixation,* University of Wisconsin Press, Madison, 1940, 206.
97. McCarty, R. E., Roles of a coupling factor for photophosphorylation in chloroplasts, *Annu. Rev. Plant Physiol.,* 30, 79, 1979.
98. Black, C. C., Photosynthetic carbon fixation in relation to net CO_2 uptake, *Annu. Rev. Plant Physiol.,* 24, 253, 1973.
99. Pate, J. S., Layzell, D. B., and McNeil, D. L., Modeling the transport and utilization of carbon and nitrogen in a nodulated legume, *Plant Physiol.,* 63, 730, 1979.
100. Döbereiner, J. and Boddey, R. M., Nitrogen fixation in association with gramineae, in. *Current Perspectives in Nitrogen Fixation,* Gibson, A. H. and Newton, W. E., Eds., Australian Academy of Sciences, Canberra, 1981, 305.
101. Bisseling, T., Van Den Bos, R. C., and Van Kammen, A., The effect of ammonium nitrate on the synthesis of nitrogenase and the concentration of leghaemoglobin in pea root nodules induced by *Rhizobium leguminosarum, Biochim. Biophys. Acta,* 539, 1, 1978.
102. Dazzo, F. B. and Sherwood, J. E., Trifoliin A: a Rhizobium recognition lectin in white clover roots, in *Chemical Taxonomy, Molecular Biology and Function of Plant Lectins,* Etzler, M. and Goldstein, I., Eds., Alan R. Liss, New York, 1983.
103. Hollaender, A., Ed., *Genetic Engineering for Nitrogen Fixation,* Plenum Press, New York, 1977, 538.
104. Collins, J., personal communication.

Chapter 3

PROTEIN SYNTHESIS

C. A. Price

TABLE OF CONTENTS

I. INTRODUCTION

Higher plants possess the genetic information to manufacture several thousand different kinds of proteins. Some of these, such as reserve proteins of seeds, are produced in very large amounts during narrow intervals of seed development. Others, such as the enzymes of mitochondrial electron transport, are present at moderate levels in every plant cell and must therefore be synthesized more or less continuously. Still others, such as repressors of transcription, are present at vanishingly low levels and are produced in response to developmental programs. Most of what we know about protein synthesis in plants has been learned from studying a handful of very abundant proteins, but there is no present reason to doubt that the synthesis of most plant proteins corresponds generally to models developed from studying the abundant ones.

A significant complication of protein synthesis in higher plants is the existence of three distinct genomes located in the nucleus, the plastids, and the mitochrondria. The nucleus contains 100,000 genes (a wild guess based on estimates of the complexity of nuclear DNA), the plastid contains 100 protein genes (another guess, but certainly between 50 and 300), and the mitochondria contain between 20 and several hundred. We can be most confident about the genetic information of the plastid because of progress made in physical mapping of its simply and extraordinarily conservative DNA. Nuclear DNA in contrast is dauntingly complex and varied. Mitochondrial DNA of higher plants is extremely puzzling; its variability in size and complexity departs strongly from the rather simple patterns shown by mitochondria of other organisms.

We shall discuss models of protein synthesis in plants (1) from the standpoint of proteins coded in the nucleus, then (2) models for the synthesis of plastid proteins, which involve events both within and without the plastid. We shall then outline the limited information concerning synthesis of mitochondrial protein and, finally, examine our understanding of the regulation of protein synthesis as it might assist the plant breeder.

II. SYNTHESIS OF NUCLEAR-CODED PROTEINS

Any description of protein synthesis has to begin with the gene. The gene itself is a sequence of DNA composed of three normally distinct regions: the 5′-flanking sequence, the transcribed portion, and the 3′-flanking sequence. An example is the sequence of a 19-kDa zein gene of maize in Figure 1.

The region between 60 bases before the start of transcription to 20 bases after typically binds RNA polymerase in *Escherichia coli;* this region is called the "promoter" (see Section V.B). Interactions between RNA polymerase and promoter sequences are thought to determine the timing and abundance of specific RNAs.

The transcription of protein genes begins at an initiation site, typically a CAT sequence, a certain distance before the start codon (normally AUG), and ends a certain number after the stop codon. In the case of many nuclear genes, there are also untranslated regions within the coding region which are called *introns.* (Introns may code for polypeptides with processing or regulatory functions.) Before the mRNA can be translated, the intron must be spliced out of the primary transcript by methods that are not well understood.

The 5′ untranslated region of the mRNA, contains a sequence of 10 nucleotides before the start of transcription, which is analogous to the "Pribnow box" of prokaryotic messages. This sequence is rich in A and T. There are other "concensus sequences" at -35 and -43 nucleotides which enable the message to bind to the 18S RNA of the 40S ribosome. The 3′ untranslated region contains an inverted repeat which may

FIGURE 1. Sequence of a gene for a 19-kDa zein protein.[1] The continuous nucleotide sequence is that from a clone in a genomic lambda library designated λZG99. Numbering begins with the first coding sequence, which is an ATG. The sequence of a homologous cDNA clone, pZ19.1, is shown above the genomic sequence. This sequence differs from that of the genomic clone by a few scattered nucleotides, as indicated, and by a large gap shown as asterisks. The sequence of a different cDNA clone, A30,[2] is shown below the genomic sequence. The beginnings and ends of each cDNA sequence are shown as "start:" and "end:". Horizontal lines show positions of putative regulatory sequences in the gene. (From Pedersen, K., Devereux, J., Wilson, D. R., Sheldon, E., and Larkins, B. A., *Cell,* 29, 1015, 1982. With permission.)

form a stem and loop and may serve as a termination signal for the ribosome, but such secondary structures have not been established by direct determination. Finally, most nuclear messages contain a site on the 3′ end for the attachment of a poly(A) tail.

Once the message leaves the nucleus, it may bind to ribosomes. It is suspected that differences among 5′-untranslated regions of messages affect how avidly they bind to ribosomes. Stated differently, messages appear to compete with one another for the opportunity to be translated, but, except to say that translational regulation occurs, there is little information on this important phenomenon among mRNAs from nuclear genes in plants.

Translation of mRNAs in the cytoplasm occurs on 80S-types of ribosomes. The important characteristic of ribosomes is not their actual S value or sedimentation coefficient, which may depart from 80S depending on the species, but that the cytoplasmic ribosomes of higher plants are typical of eukaryotic ribosomes generally with respect to the initial amino acid (always methionine), initiation factors, and elongation factors. A corollary is that cytoplasmic protein synthesis is sensitive to the antibiotic cycloheximide and insensitive to D-threo-chloramphenicol and streptomycin.

A. Post-Translational Modifications

Many hydrophilic proteins are synthesized, fold into a comfortable tertiary structure, and there they are — mature and functional. For others, maturity does not come until the proteins undergo modifications to their covalent structure. This may entail nothing more than the condensation of two cysteines to form a disulfide bridge or it may involve phosphorylation, glycosylation, or condensation with fatty acids. Finally the polypeptide chain may be cleaved.

B. Transport

The synthesis of proteins located within or beyond a membrane entails a special set of problems regarding how to move a polypeptide from the essentially aqueous environment of the cytoplasm into or across a hydrophobic layer. The problem has been solved in two ways, by cotranslational processing and by posttranslational processing.

In cotranslational processing the initial dozen or so amino acids of the nascent polypeptide comprise a "signal peptide". This sequence binds to small ribonucleoproteins which, acting like seeing-eye dogs, lead the polysome to the outer surface of the endoplasmic reticulum. The signal peptide is drawn through pores in the endoplasmic reticulum followed by the remainder of the polypeptide. A "signal peptidase" on the interior of the lumen side of the endoplasmic reticulum then removes the signal peptide in a single proteolytic event and the remainder of the polypeptide continues on its way (Figure 2).

One can show in such cases that the polypeptide translated from the mRNA in the absence of endoplasmic reticulum is longer by the length of the signal polypeptide than the mature protein. The processing of seed proteins, such as legumin,[4] follows such a path, but in the case of seed proteins there are additional processing events: the translation product, shortened by the action of the signal peptidase, proceeds to the Golgi bodies, where sugars are added. The glycosylated protein is then cleaved into smaller subunits and eventually deposited into protein bodies of the endosperm or cotyledon (Figure 3).

Because removal of the signal peptide, glycosylation, and further cleavage successively decrease, increase, and again decrease the molecular weight of the polypeptides, the maturation of seed proteins seen by electrophoretic patterns is very complicated (Figure 4). It can, however, be analyzed by a combination of pulse labeling, subcellular fractionation, and electrophoresis.

The synthesis of membrane proteins and exported proteins follows similar courses, but in many cases the leader sequence, which facilitates attachment of the polysome to the endoplasmic reticulum and transport across the membrane, is not subsequently removed.

C. Post-Translational Processing

In the synthesis of chloroplast and mitochondrial proteins in the cytoplasm, the mRNA is translated on free ribosomes (rather than of ribosomes attached to membranes, as in the example above), the complete polypeptide migrates to the surface of the organelle, is somehow recognized, and is transported across the envelope or outer membrane. Essential to this transport is an N-terminal transit peptide, similar in function to the signal peptide. The transit peptide is cleaved in one or more steps within the organelle by specific proteases.

Post-translational processing differs from cotranslational processing in that the ribosomes do not bind to the membrane and proteolysis occurs after the polypeptide has left the ribosome, rather than during translation.

D. Multigene Families

Most of the genes that have been isolated from plant nuclear DNA occur as members of multigene families located in a dozen or more distinct sequences. The members of a

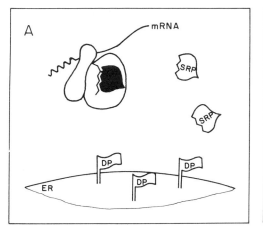
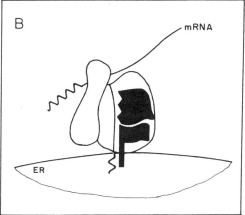
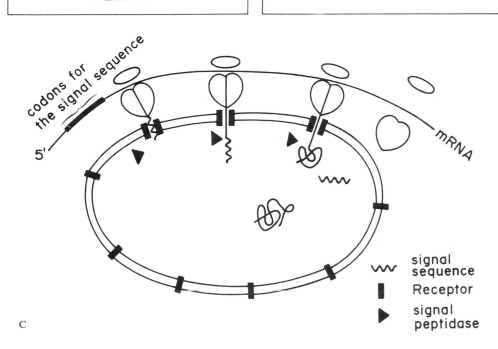

FIGURE 2.   Mechanism of cotranslational processing. The figure depicts events occurring in the ER-associated translation of proteins. (A) mRNA bearing a signal sequence binds to a ribosome. When the signal sequence has been translated, a signal recognition protein (SRR) binds to the ribosome and stops translation. (B) Translation remains arrested until the ribosome complex contacts a receptor in the endoplasmic reticulum (ER) called a docking protein (DP). The nascent chain then crosses the ER by an unknown process and translation resumes. (C) When the signal sequence emerges into the lumen of the ER it becomes susceptible to proteolytic cleavage by the signal peptidase. The remainder of the polypeptide is eventually released into the lumen.

family can be detected by the method of DNA gel transfer and homologous hybridization.[6] Specifically nuclear DNA is cut with a restriction enzyme, the fragments separated by electrophoresis, then transferred and fixed to nitrocellulose paper. The transferred fragments are hybridized with DNA from an isolated gene that has been made radioactive. If the gene occurred only once in the genome or in multiple copies containing identical sequences on either side, the radioactive probe would "light up" a

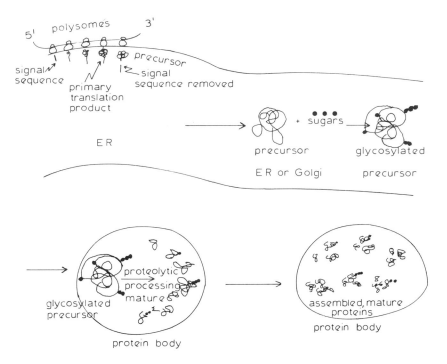

FIGURE 3.   Biosynthesis of seed storage protein.[4] The primary translation product is produced in the endoplasmic reticulum (ER) by cotranslational processing. Glycosylation takes
place in the ER or in the Golgi bodies. The polypeptides are then transported to protein bodies
where proteolytic cleavage and assembly into mature proteins occurs.

single band on the gel. What is almost invariably observed, however, is that the problem hybridizes to multiple locations on the nitrocellulose replica of the gele (Figure 5).
In some instances the gene copies are adjacent on the chromosome and tandemly ar-
ranged; in other cases they are widely separated. Examination of these gene copies in
several instances (e.g., genes for storage proteins, the chl a/b-binding protein, and the
small subunit of RuBPC) has shown that the different copies are similar, but not identical to one another. Some of the replacements in nucleotides are "silent", i.e., code
for the same sequence of amino acids. In other cases there are small alterations in the
primary structure of the protein. Differences have also been detected in the 5'- and 3'-
flanking sequences (See Section V. B).

The existence of multiple copies of genes for major plant proteins (we know little of
any minor proteins) has led to the use of the terms "family" and "tribe" to describe
genes that are more or less conserved. The discovery of multigene families at the molecular level comes as no surprise to plant breeders, who have been contending for
years with multigene and quantitative inheritance of important characters.

III. PLASTID-CODED PROTEINS

Plastids of higher plants contain DNA of 121 to 161 kbp.[7] Plastid DNA is circular
and present as a few molecules (3 to 5) in proplastids or many (50 to 200) in mature
chloroplasts. A map of a plastid genome typical of higher plants is shown in Figure 6.
A characteristic feature of most plastid DNAs is an inverted repeat, containing the
genes for plastid rRNAs and some of the tRNAs. A portion of the *Phaseolus* family
has a single rDNA region (Figure 7). Table 1 lists the proteins known to be synthesized
in plastids together with those whose gene locations have been mapped.

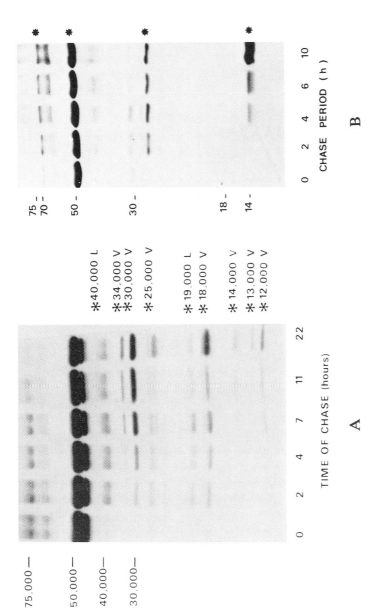

FIGURE 4. Electrophoretic analysis of posttranslational modification of a seed-storage protein. (A) Change in the labeling pattern of polypeptides during pulse-chase experiments. Pea coytledons were labeled for 90 min with [¹⁴C]amino acids and transferred to a nutrient medium at 0 time. Protein-body proteins were separated on an immunoaffinity column and the polypeptides separated by SDS-gel electrophoresis. The labeled polypeptides were then visualized by fluorography. Asterisks denote those legumin (L) and vicilin (V) polypeptides that increase during the chase and therefore represent posttranslational products. (B) Glycosylation of protein-body polypeptides during the chase period. Pea cotyledons were labeled for 2 hr with [¹⁴C]glucosamine and chased for the indicated time in cold glucosamine. The polypeptides marked with asterisks correspond to the major glyco-polypeptides of vicilin. The data show that all of the glycopeptides, with the possible exception of the 50-kDa component, are posttranslational products. (From Chrispeels. M. J., Higgins, J. J. V., and Spencer, D., *J. Cell Biol.*, 93, 306, 1982. With permission.)

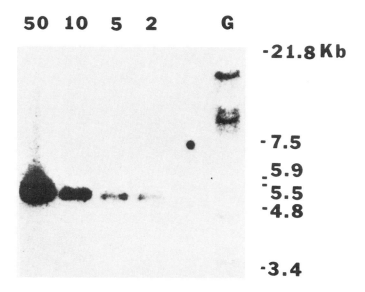

FIGURE 5.   Identification of multiple gene families by "Southern" hybridization. Maize DNA was restricted with EcoR1, separated by electrophoresis, and transferred to nitrocellulose paper by the method of Southern.[6] DNA prepared from a cDNA clone (pZ19.1) was used to probe the maize DNA. The right-hand lane, designated "G", contains DNA corresponding to 10^6 copies of the maize genome. Hybridization between maize DNA and the cDNA probe is detected at seven locations between 3 and 20 kb. The lanes to the left contain cDNA corresponding to 50, 10, 5, and 2×10^6 copies and thus serve as internal quantitative standards. Compared to these internal standards, the three upper bands in lane G contain sequences that appear to be present in three to five copies per genome, whereas the faint, lower bands correspond to single-copy sequences. (From Pedersen, K., Devereux, J., Wilson, D. R., Sheldon, E., and Larkins, B. A., *Cell,* 29, 1015, 1982. With permission.)

The plastid genome codes for all of the rRNAs and tRNAs of the plastid and for proteins of every compartment of the plastid: envelope, thylakoid, and stroma. Plastid genes account for about one half of the thylakoid proteins. Among the enzymes of the Calvin-Benson cycle of carbon reduction, only ribulose-bisphosphate carboxylase (RuBPC) has yielded to genetic analysis: the large subunit (LS) is coded on plastid DNA and the small subunit (SS) is coded in the nucleus. RuBPC can be thought of therefore as a hybrid oligomer.

RuBPC and coupling factor $(CF)_1$ were the first proteins to be identified as products of chloroplast protein synthesis [8,9] and both are hybrid oligomers. This pattern, however, is not a general one, as a number of proteins are made entirely within the chloroplast (cf., Table 1).

The protein synthetic system in plastids is similar to that in prokaryotes. The similarities include the size of ribosomes, sensitivity to inhibitors (such as D-threo-chloramphenicol), RNA sequences at ribosome-binding sites, properties of initiation and elongation factors, and the probable occurrence of polycistronic messages. In addition, a number of plastid genes show a sufficiently strong homology with the corresponding bacterial genes that hybridization with probes constructed from bacterial genes provides one of the most powerful means for locating genes on plastid DNA.

Plastid genomes also have some eukaryotic characters, including introns in rRNA genes. Introns have also been detected in protein genes in plastids of algae,[23,24] but not yet in those of higher plants.

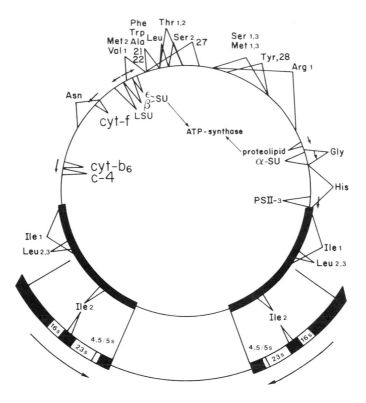

FIGURE 6. Map of the plastid genome from spinach. The spinach plastid genome is a closed circle. The large arrow heads represent recognition sites for the restriction endonuclease Sal 1. DNA fragments generated by restriction with Sal 1 are numbered in order of decreasing size. The regions containing specific genes are shown by light and heavy lines. Small arrows show the direction of transcription of ribosomal RNA operons. (From Bohnert, H. J., Crouse, E. J., and Schmitt, J. M., *Encyclopedia of Plant Physiology*, NS, Vol. 14B, Parthier, B. and Boulter, D., Eds., Springer, Berlin, 1982. With permission.)

FIGURE 7. Three types of plastid genomes. Each of the three types of plastid genomes are closed circles of DNA. Arrows show repeat sequences and directions of transcription of ribosomal RNA operons. Closed circles represent 16S rRNA genes; open circles represent 23 S rRNA genes. (From Bohnert, H. J., Crouse, E. J., and Schmitt, J. M., *Encyclopedia of Plant Physiology*, NS, Vol. 14B, Parthier, B. and Boulter, D., Eds., Springer, Berlin, 1982. With permission.)

Table 1
PLASTID GENES AND TRANSLATION PRODUCTS

Protein or polypeptide	Translation product	Gene mapped
LS or RuBPC	Pea[8]	*Chlamydomonas*[11]
Coupling factor subunits		
Alpha subunit	Spinach[9]	Spinach[12]
Beta	Spinach[9]	Spinach[12]
Epsilon	Spinach[9]	Spinach[12]
I	Spinach[13]	Spinach[12]
III	Spinach[13]	Spinach[12]
Elongation factors		
G	Spinach[14]	—
Tu	Spinach[14]	Spinach[15]
Cytochrome b559	Spinach[16]	—
Cytochrome f	Spinach[17]	Spinach[15]
Cytochrome b563	Spinach[15]	Spinach[15]
tmpA (=32 kDa protein)	Pea[18]	Maize[19]
P700-chl a protein	Spinach[16,18,20]	Spinach[16]
Rieske protein	Spinach[15]	Spinach[15]
Ribosomal proteins	Spinach[21]	*Chlamydomonas*[22]

Note: The species are listed in which the protein was first identified as a
plastid translation product or the gene mapped. The identification
of a protein as a plastid translation product has frequently led to
the localization of the gene on the plastid genome.

From Price, C. A., Miller, M. E., and Reardon, E. M., *The New Frontiers in Plant Biochemistry*, Akazawa, T., Asahi, T., and Imaseki, H.,
Eds., Nijhoff/Junk, The Hague, 1983, 107. With permission.

Some plastid proteins (e.g., cytochrome *f*) are made as precursors. It is possible, but
not proven, that some integral proteins of thylakoids are made by a kind of cotranslational processing.

IV. MITOCHONDRION-CODED PROTEINS

The mitochondrial genomes of humans, rats, and yeast have been completely
sequenced[25,26] and are found to code for rRNAs, tRNAs, and virtually the same handful of polypeptides of the inner mitochondrial membrane. In contrast to the tidy state
of molecular genetics of these mitochondria, the situation in plants is chaotic: genome
size ranges from a minimum of 200 kbp (maize) to a maximum of 2400 kbp (water-
melon). The main mitochondrial DNA is probably circular, but there are other bits
and pieces which are mostly linear.

The maize genome probably codes for many proteins in addition to the few coded
by mitochondrial DNA in yeast, but progress in identifying genes in plant mitochondria has been very slow. Apart from the purely logistic problems of dealing with DNA
of this size and the relatively few laboratories working on the problem, the interpretation of hybridization data is complicated by evidence[27] that mitochondrial DNA includes fragments of plastid genes, truncated and slightly scrambled, and almost certainly not functional. It is as if plant mitochondria are a kind of genetic garbage heap
having stably incorporated flotsam from other genetic systems of the plant cell.

As if the main band of mitochondrial DNA were not sufficiently complex, there is a
tantalizing correlation in a number of plants between male sterility and the occurrence
of one or two small (5- to 6-kbp) fragments of DNA (cf., Reference 28). In maize these

are called S1 and S2. These fragments appear not to be transcribed, but are homologous with sequences in the main band of normal, male-fertile mitochondrial DNA. S1 and S2 act in some ways as if they were transposons that can move out of the main-band DNA and thus render a gene inoperative. There is thus far no correlation between the physiological defect observed in male-sterile pollen and the changes detected in mitochondrial DNA.

V. REGULATION OF GENE EXPRESSION IN PLANTS

It is obvious to anyone who has observed plants that different characters are expressed differentially during development. Potato plastids become chloroplasts in leaves and amyloplasts in tubers; in *Petunia,* anthocyanins are formed in the petals but not in roots; in pea, the axillary buds first differentiate into tendrils, whereas later ones become flowers. Various kinds of direct genetic evidence in each of these examples demonstrate that we are witnessing the expression of different genes during the course of plant development.

There are at least five conceivable levels at which regulation of gene expression in plants could occur: activation or inactivation of genes by transposons; control of transcription; control of post-transcriptional processing; control of translation; and control of post-translational processing. Of these, there is documentation in plants only for regulation at the levels of transcription and translation. We shall discuss transcriptional regulation, the nature of promoters (which are the sites at which this kind of regulation occurs), some possibilities for post-transcriptional regulation, and finally translational regulation.

A. Transcription

It has become evident only recently that many cases of differential gene expression correspond to differential transcription of the gene. Two examples will suffice.

1.   Seed proteins are synthesized during a narrow interval between anthesis and maturation (cf., Reference 29 and Figure 8). Immediately preceding this interval, one can detect the formation of the corresponding mRNA. Using a seed protein gene as a very sensitive and quantitative probe, the transcripts are seen to form only during seed development and never again during the life of the plant. A seeming exception is the seed lectin of soybean, which also appears in root hairs. The root-hair lectin, however, is coded by a gene with the same coding sequence as that of the seed lectin, but a different promoter. The two genes thus code for the same protein but respond to different developmental signals.
2.   Bundle-sheath cells of C4 plants contain an active RuBPC, whereas cells from the same plants are substantially devoid of RuBPC.[30] The plastid DNAs of the two kinds of cells are indistinguishable, but mRNAs for LS can be detected only in bundle-sheath cells.

We can conclude therefore that the synthesis of seed proteins and of RuBPC in these two examples are the results of transcriptional regulation.

B. Promoters

As described above, RNA polymerase attaches to the promoter of a gene, usually located in the 5′-flanking region, but sometimes extending into the coding region. Eukaryotic promoters typically contain a CATT sequence at about -80 from the start site and a TATAATA sequence at about -30. Prokaryotic promoters, such as one would expect to find in plastid genes, typically contain a TATAAT ("Pribnow box") at -10,

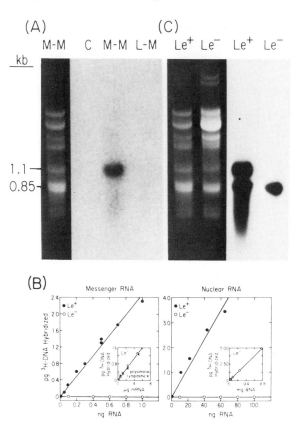

FIGURE 8.   Developmental regulation of synthesis of a soybean seed protein. Poly (A) + RNAs were extracted from soybean tissues, separated by electrophoresis, and transferred to nitrocellulose filters ("northern blots"). A cDNA probe to soybean lectin was used to probe the transferred mRNAs from cotyledons (C), embryos from mid-maturation phase (M-M), and late-maturation phase (L-M). The lane at the far left is an RNA gel stained with ethidium bromide. The cDNA probe hybridizes with RNA corresponding to a prominent RNA species of about 1.1 kb, but only in samples from mid-maturation embryos. (From Goldberg, R. B., Hoschek, G., and Vodkin, L. O., *Cell*, 33, 465, 1983. With permission.)

a TTGACA at -35, and an AT-rich region at -43. Beyond these generalizations, we know very little about what signals are responsible for transcriptional regulation.

We can nonetheless exploit the existence of promoters in genetic engineering. The Gent-Cologne group (cf., Reference 31), for example, transformed tobacco with a plasmid bearing the promoter for SS fused to the coding sequence of a gene for antibiotic resistance. The resulting tobacco calli were resistant to the antibiotic only in the light. In another application, a phaseolin coding sequence was fused to the promoter for octopine synthetase and employed to transform sunflower.[32] The transformed plants produced phaseolin.

These extraordinary experiments mean that we can hope to control the expression of native or foreign genes almost at will. We can imagine that distinct promoters exist that respond to all manner of external signals, such as light, temperature, nutrient levels, photoperiod, and pathogenic elicitors; whereas others are keyed to internal signals such as developmental stages, anatomical location, and metabolite levels. The promoters will be isolated from genes that respond to a signal of specific interest. They would then be fused to YFGs (your favorite genes), cloned into convenient shuttle vectors, and employed for transformation of a target plant. Such a program appears feasible even though we are presently ignorant of the identities of the repressors and derepressors that presumably act as the immediate agents for regulating transcription in plants.

C. Post-Transcriptional Processing

As noted earlier, many eukaryotic genes contain untranslated interruptions called introns and the excision of introns and splicing together of exons are essential steps in information transfer. In some systems a single exon may be spliced to two or more distinct exons in different locations to make two distinct proteins. The operational concept here is that an exon may code for specific functional domain of an enzyme and that the same domain may be employed in two or more locations.

Although no examples of regulation at the level of splicing have yet been found with plant material, we should be prepared for the possibility. The first intron of the SS gene, for example, occurs almost exactly at the end of the sequence coding for the transit peptide. It is conceivable, for example, that the gene for this transit peptide might be employed with a different gene requiring the same transit sequence.

D. Translation

Lemna grown in the dark with intermittent red light accumulate mRNA for the chlorophyll *a*/b binding protein, but do not synthesize the protein.[33] Similarly proplastids of *Euglena* contain normal levels of mRNA for LS, but do not synthesize the polypeptide at significant rates.[34] There may be many more instances where the expression of a gene is arrested at some level subsequent to transcription. Apart from a general observations that different mRNAs can compete with one another for ribosome binding, we know very little about the mechanisms of regulation at this level.

VI. CONCLUSIONS

We have gained a broad understanding of how genes code for proteins in nuclei, plastids, and mitochondria, and a glimpse at the sequential events of transcription, translation, transport and assembly. The development of means to identify and isolate individual genes, the ability to transform higher plants with exogenous genes, and the knowledge of mechanisms for regulation of gene expression offer the promise of controlling the composition and development of plants through genetic engineering.

REFERENCES

1. Pedersen, K., Devereux, J., Wilson, D. R., Sheldon, E., and Larkins, B. A., Cloning and sequence analysis reveal structural variation among related zein genes in maize, *Cell*, 29, 1015, 1982.
2. Geraghty, D., Peifer, M. A., Rubenstein, I., and Messing, J., The primary structure of a plant storage protein:zein, *Nucl. Acids Res.*, 9, 5163, 1981.
3. Leader, D. P., Protein biosynthesis on membrane-bound ribosomes, in *DNA Makes RNA Makes Protein*, Hunt, T., Prentis, S., and Tooze, J., Eds., Elsevier, Amsterdam, 1983, 257.
4. Higgins, T. J. V., Chandler, P. M., Spencer, D., Chrispeels, M. J., and Zurawski, G., The synthesis, processing and primary structure of pea seed lectin, in *Structure and Function of Plant Genomes*, Ciferri O., and Dure, L., Eds., Plenum Press, New York, 1983, 93.
5. Chrispeels, M. J., Higgins, J. J. V., and Spencer, D., Assembly of storage protein oligomers in the endoplasmic reticulum, *J. Cell Biol.*, 93, 306, 1982.
6. Southern, E. M., Detection of specific sequences among DNA fragments separated by gel electrophoresis, *J. Mol. Biol.*, 98, 503, 1975.
7. Bohnert, H. J., Crouse, E. J., and Schmitt, J. M., Organization and expression of plastid genomes, in *Encyclopedia of Plant Physiology*, NS, 14B, Parthier B. and Boulter D., Eds., Springer, Berlin, 1982, 475.

8. Blair, E. J. and Ellis, R. J., Protein synthesis in chloroplasts. I. Light-driven synthesis of the large subunit of fraction I protein by isolated chloroplasts, *Biochim. Biophys. Acta,* 319, 223, 1973.

9. Mendiola-Morgenthaler, L. M., Morgenthaler, J.-J., and Price, C. A., Synthesis of coupling factor CF1 protein by isolated spinach chloroplasts, *FEBS Lett.,* 62, 96, 1976.

10. Price, C. A., Miller, M. E., and Reardon, E. M., Protein synthesis in chloroplasts, in *The New Frontiers in Plant Biochemistry,* Akazawa, T., Asahi, T., and Imaseki, H., Eds., Nijhoff/Junk, The Hague, 1983, 107.

11. Rochaix, J.-D. and Darlix, A.-L., Composite structure of the chloroplast 23S ribosomal RNA genes of *Chlamydomonas reinhardii.* Evolutionary and functional implications, *J. Mol. Biol.,* 159, 383, 1982.

12. Westhoff, P., Nelson, N., Buenemann, H., and Herrmann, R. G., Localization of genes for coupling factor subunits on the spinach plastid chromosome, *Curr. Genet.,* 4, 109, 1981.

13. Nelson, N., Nelson, H., and Schatz, G., Biosynthesis and assembly of the proton-translocating adenosine triphosphatase complex from chloroplasts, *Proc. Natl. Acad. Sci. U.S.A.,* 77, 1361, 1982.

14. Ciferri, O., De Pasquale, G., and Tiboni, O., Chloroplast elongation factors are synthesized in the chloroplast, *Eur. J. Biochem.,* 102, 331, 1980.

15. Herrmann, R. G., Alt, J., Bisanz, C., Hauska, G., Nelson, N., Westhoff, P., and Winter, P., Function and organization of dicotyledon plastid chromosomes, *Structure and Function of Plant Genomes,* Ciferri, O. and Dure, L., Eds., Plenum Press, New York, 1983.

16. Zielinski, R. E. and Price, C. A., Synthesis of thylakoid membrane proteins by chloroplasts isolated from spinach: cytochrome b559 and P700-chlorophyll a-protein, *J. Cell Biol.,* 85, 435, 1980.

17. Doherty, A. and Gray, J. C., Synthesis of cytochrome f by isolated pea chloroplasts, *Eur. J. Biochem.,* 98, 89, 1979.

18. Eaglesham, A. R. J. and Ellis, R. J., Protein synthesis in chloroplasts. II. Light-driven synthesis of membrane proteins by isolated pea chloroplasts, *Biochim. Biophys. Acta,* 335, 396, 1974.

19. Bogorad, L., Chloroplasts, *J. Cell Biol.,* 91, 2565, 1981.

20. Cederblad, A. V. and Vasconcelos, A. C., PSI reaction center protein — site of synthesis of constituent polypeptides, *Plant Physiol.,* 61 (Suppl.), 83, 1978.

21. Mache, R., Dorne, A. M., Lescure, A. M., and Eneas-Filho, J., Biogenesis of plastid ribosomes from spinach: site of synthesis and assembly, in *Structure and Function of Plant Genomes,* Ciferri, O. and Dure, L., Eds., Plenum Press, New York, 1983.

22. Bogorad, L., Genes for chloroplast ribosomal RNAs and ribosomal proteins: gene dispersal in eukaryotic genomes, in *International Cell Biology,* Brinkley, B. R. and Porter, K. R., Eds., Rockefeller University Press, New York, 1977, 175.

23. Stiegler, G. L., Matthews, H. M., Jr., Bingham, S. E., and Hallick R. B., Location on the *Euglena gracilis* chloroplast DNA of genes for the large subunit of ribulose-1,5-bisphosphate carboxylase and two thylakoid membrane polypeptides, *Fed. Proc. Fed. Am. Soc. Exp. Biol.,* 41, 758, 1982.

24. Rochaix, J.-D., Erickson, J., Dron, M., Schneider, M., and Vallet, J. M., Chloroplast genes and transformation in *Chlamydomonas* reinhardii, in *Advances in Photosynthesis Research,* Vol. 4, Sybesma, C., Ed., Martinus Nijhoff/Dr. W. Junk, The Hague, 1984, 491.

25. Bernardi, G., The petite mutation in yeast, in *DNA Makes RNA Makes Protein,* Hunt, T., Prentis, S., and Tooze, J., Eds., Elsevier, Amsterdam, 1983, 67.

26. Attardi, G., Organization and expression of the mammalian mitochondrial genome: a lesson in economy, in *DNA Makes RNA Makes Protein,* Hunt, T., Prentis, S., and Tooze, J., Eds., Elsevier, Amsterdam, 1983, 85.

27. Stern, A. I. and Lonsdale, D. M., Mitochondrial and chloroplast genomes of maize have a 12-kilobase DNA sequence in common, *Nature (London),* 299, 698, 1982.

28. Levings, C. S., Plasmid-like DNAs in the mitochondria of higher plants, in *Structure and Function of Plant Genomes,* Ciferri, O. and Dure, L., Eds., Plenum Press, New York, 1983.

29. Goldberg, R. B., Hoschek, G., and Vodkin, L. O., An insertion sequence blocks the expression of a soybean lectin, *Cell,* 33, 465, 1983.

30. Link, G., Coen, D. M., and Bogorad, L., Differential expression of the gene for the large subunit of ribulose bisphosphate carboxylase in maize leaf cell types, *Cell,* 15, 725, 1978.

31. Caplan, A., Herrera-Estrella, L., Inze, O., van Haute, E., van Montagu, M., Schell, J., and Zambryski, P., Introduction of genetic material into plant cells, *Science,* 222, 815, 1983.

32. Murai, N., Sutton, D. W., Murray, M. G., Slightom, J. L., Merlo, D. J., Reichert, N. J., Gengupta-Gopalan, C., Stock, C. A., Barker, R. F., Kemp, J. D., and Hall, T. C., Phaseolin gene from bean is expressed after transfer to sunflower via tumor-inducing plasmid vectors, *Science,* 222, 476, 1983.

33. Slovin, J. P. and Tobin, E. M., Synthesis and turnover of the light-harvesting chlorophyll a/b protein in *Lemna gibba* grown with intermittent red light: possible translational control, *Planta,* 154, 465, 1982.

34. Miller, M. E., Jurgenson, J. E., Reardon, E. M., and Price, C. A., Plastid translation *in organello* and *in vitro* during light-induced development in *Euglena, J. Biol. Chem,* 258, 14478, 1983.

Chapter 4

UPTAKE, TRANSLOCATION, AND REDUCTION OF NITRATE

W. A. Jackson, W. L. Pan, R. H. Moll, and E. J. Kamprath

TABLE OF CONTENTS

I. INTRODUCTION

Intraspecific differences in growth response of nonnodulating plants to nitrogen supply have been recognized for a number of years. With maize, for example, genotypic differences in grain yield response to nitrogen fertilization were reported 50 years ago.[1,2] An example of differential response in above-ground biomass of maize to increasing nitrogen supply [3] is shown in Figure 1. Population hybrids of "Jarvis × Indian Chief" after eight cycles of full-sib family selection and eight cycles of reciprocal recurrent selection for interpopulation yield performance were compared at three levels of applied nitrogen. Both selection procedures increased biomass production at low levels of nitrogen supply, but the patterns of response to increasing nitrogen differed. Reciprocal recurrent selection resulted in a population that gave as much biomass at 168 kg N ha^{-1} as the full-sib family selection at 280 kg N ha^{-1}. The efficiency of the two improved materials to utilize soil and/or applied nitrogen for maximal dry weight accumulation differed appreciably. Such differences must result from a differential efficiency in the acquisition of nitrogen from the soil solution, a differential efficiency in utilization of the absorbed nitrogen for dry matter (including grain) production, or a combination of the two.

It has also been shown that significant variation occurs in the total amount of nitrogen taken up by maize genotypes under field conditions[3-13] as well as by other species.[14-16] In addition to genotypic differences in total nitrogen absorbed by maize over the growing season, variation can occur in the proportion absorbed after the onset of reproductive growth.[3,10-12]

An example of the range in nitrogen uptake that can occur is shown in Figure 2 for eight experimental maize hybrids.[12] All hybrids (single crosses between unselected inbred lines, S > 18 generation) from "Jarvis Golden Prolific" and "Indian Chief" were grown to physiological maturity on a Dothan loamy sand with a low supply of fertilizer nitrogen (2.47 g N plant^{-1}). Although all of the hybrids accumulated more nitrogen in their above-ground biomass than was applied as fertilizer, indicating effective uptake from residual soil nitrogen sources, the nitrogen absorbed by the hybrids varied significantly. For six of the eight, a linear relationship (r = 0.93) was evident between the above-ground biomass and the nitrogen in that biomass. However, genetic diversity in this trait is indicated by one hybrid having significantly more, and another significantly less, biomass per unit of nitrogen in the tissue. This genetic diversity in nitrogen uptake, if combined with traits that foster efficient utilization of the absorbed nitrogen for grain or dry matter accumulation, should provide opportunities for enhancing the efficiency of fertilizer nitrogen use.

In most arable soils, nitrification results in nitrate being the dominant inorganic nitrogen species, although the relative proportion of ammonium and nitrate at absorbing surfaces of roots may vary substantially with fertilization regimes and soil conditions — including temperature, acidity and soil transport properties.[17] Because of the predominance of nitrate in arable soils, and the restricted transport in soil of ammonium to root surfaces as compared with nitrate,[18] it seems reasonable to assume that nitrate is the ion which is predominately absorbed by the roots of crop species growing on arable soils. The precise proportion taken up as each ionic species, however, is nearly impossible to ascertain in any soil-plant system. Nevertheless, in view of the indirect evidence that indicates nitrate as the dominant absorbed form, developing an understanding of the nature of regulation of nitrate uptake by root systems and how the subsequent nitrate assimilation processes are influenced by the uptake process may be helpful in attempts to enhance efficiency in nitrogen use by crop plants.[15,19]

In the following sections, we describe some factors involved in nitrate acquisition by root systems, the partitioning of entering nitrate to reduction in roots and translocation

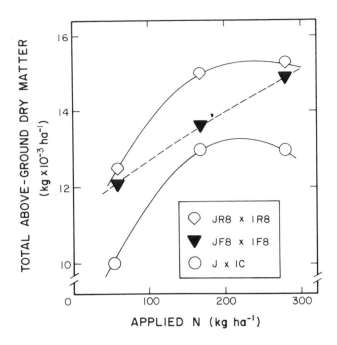

FIGURE 1.   Growth responses at physiological maturity of Jarvis × Indian Chief maize population hybrids to applied nitrogen on a Dothan loamy sand (Typic Plinthic Paleudult) after eight cycles of full-sib family selection (JF8 × IF8) or eight cycles of recurrent selection (JR8 × IR8) for interpopulation yield performance. J × IC is the original hybrid.[3]

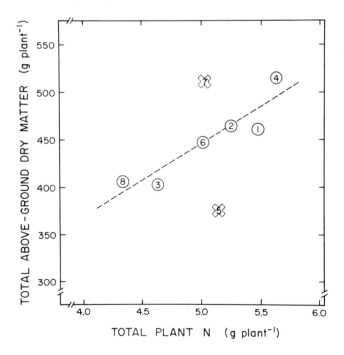

FIGURE 2.   Relationship of total above-ground dry matter and nitrogen at physiology maturity of eight experimental maize hybrids on a Typic Plinthic Paleudult.[12] The dotted line is the regression (Y − 62.12 + 76.65X, r = 0.926) for the six hybrids shown in the open circles.

to shoots, and the relatively little-studied relationship between nitrate translocation and reduction in leaf cells. We conclude by showing briefly how relatively simple measurements of dry matter and nitrogen may be used to describe some of the traits which may indicate the effectiveness with which nitrogen assimilation processes contribute to dry matter or grain production, and how these measurements may be utilized to advantage in programs of genetic selection and in the development of specific genotype-management systems.

II. EFFICIENCY IN NITRATE UPTAKE BY HIGHER PLANTS

A. Proliferation of the Root System

The total quantity of nitrogen absorbed is a function of the amount of root tissue capable of absorption, the absorption rate of each segment of the root system, and time. Both root parameters are affected markedly by the plant growth rate. Genotypic differences in developmental patterns of roots are evident in a number of species.[20-27] The evidence indicates that selection and development of particular kinds of root systems to facilitate adaptation to certain stress condition can be accomplished. An example is the variation in root parameters among sorghum genotypes,[28] one of which has the capability to produce relatively large root length densities ($cm_{root}cm_{soil}^{-3}$) in the lower regimes of the soil profile at times when maximal water removal from these regions is required.[29,30]

Even though substantial genetic diversity in root proliferation occurs, chemical and physical attributes of the soil also can profoundly modify rooting patterns.[31] Hence, genetic manipulations of root proliferation patterns of a given species to optimize water and nutrient acquisition have to be evaluated for relatively specific soil conditions. Utilizing genetic diversity to enhance nitrogen acquisition has the additional complication that proliferation is altered substantially by the nitrogen supply and its location within the root zone. The extent to which this response varies genetically is not known. Lateral initiation and extension can be markedly stimulated in localized regions exposed to a high nitrate supply when the remainder of the root system is exposed to low supplies.[32-34] Moreover, in a uniform environment, the ambient nitrogen concentration for maximal lateral initiation may not be identical with that for maximal axis or lateral elongation.[35,36] These marked effects of the amount and distribution of nitrogen in the soil profile on root proliferation and morphological patterns, and the relatively high mobility of nitrate in the soil, indicate that the success of programs to alter root proliferation by genetic manipulation to enhance nitrogen acquisition would be uncertain. Yet there are conditions where distinct advantages may be obtained. For example sandy soils of the coastal plain of the southeastern U.S. have E horizons whose physical and chemical attributes tend to restrict root proliferation. At the same time, nitrate can be leached below this layer and thereby become positionally unavailable. Genotypes with roots that can move through this layer would have access to additional nitrogen as well as water supplies.

B. The Relationship Between Ambient Concentration and Nitrate Uptake Rate

Frequent additions of nitrogen at specific fractions of the amount required for maximal growth rates has shown, at suboptimal nitrogen supplies, strong linear relationships between the amount of nitrogen added per unit time, the relative growth rate of the plants, and the percentage nitrogen in the plants.[37-41] Throughout the suboptimal range, essentially all the nitrogen was removed from the solutions, indicating a high efficiency of absorption at low ambient concentrations. Furthermore, high growth rates have been maintained at low ambient nitrogen concentration in flowing nutrient solution experiments. Wheat seedling growth was sustained at 5 μM nitrate, without the appearance of visually detectable nitrogen deficiency symptoms, provided solution flow rates were adequate to maintain a constant ambient concentration of nitrate.[42] Maximal growth of perennial ryegrass occurred at 1.4 mM nitrate, but only a 10% decrease in shoot dry weight was observed in plants grown in an ambient nitrate concentration 100-fold lower.[43] In contrast, growth responses were observed in corn seedlings to nitrate concentrations in excess of 7.5 mM, when a slantboard technique was used.[36] Overall, these experiments indicate that nitrate may be absorbed efficiently at low ambient concentration but the concentration at which optimum growth rates are maintained may vary with species and/or environmental conditions.

Nitrate uptake rates within a specific concentration range show progressive saturation curves,[44-54] as is characteristic of the uptake of other inorganic ions. Such curves can be quantified by the maximal rate attained where ambient concentration is not limiting the system (Vmax), and the by the apparent Km, the concentration at which half the maximal rate occurs. The latter indicates the affinity of the ambient ions for the transport system. Because nitrate uptake rates are so closely associated with plant growth,[55-57] energy supply to the root system,[58] and to the nitrogen status of the plants at the time the measurements are made,[50,52,53,59-63] the Vmax and Km values obtained in short-time experiments reflect only the status of the plants at the time of measurement[50] and are not always an inherent attribute of the plants.[64] This is illustrated by an investigation of five genotypes differing in their natural habitats from nutrient-rich to nutrient-poor conditions.[49] Variations in Vmax and Km values for nitrate uptake were found among the genotypes at various times during their growth at high and low nutrient supplies but the values did not correspond to those expected from their adaptations to the natural environment (a low Km for nutrient-poor conditions and a high Vmax for nutrient-rich conditions). On the other hand, agreement between natural adaptation and short-term nitrate uptake has been observed by Bloom and Chapin.[65] They showed that a barley variety adapted to warm soils (where the proportion of nitrate to ammonium in the soil would be large) had a substantially larger Vmax for nitrate uptake at 15°C than a barley variety adapted for growth in cold soils, where ammonium would be expected to dominate the soil supply. In contrast, the cold-adapted variety was more effective in ammonium uptake at 5°C.

Differences in nitrate uptake among genotypes within a species have been demonstrated frequently.[14,15,49,66-72] In some instances, the differences could not be accounted for fully by variations in root mass[8] or total root length per plant.[71,73] A twofold range in nitrate uptake per unit root volume has been observed with roots of a group of young, dark-grown, decapitated inbred maize seedlings (Table 1). Within the group, however, there were differing capabilities for nitrate translocation, reduction, and accumulation by the root tissue. Because genetic differences in nitrate partitioning after absorption could modify the uptake rates (see below), it is not possible to state conclusively that the differences in the uptake rates were a result of a specific genetic lesion in that process.

Taken as a whole, the data indicate that genetic adjustments in nitrate uptake (and ammonium uptake as well) have occurred during adaptation to specific soil and environmental conditions, and that genetic diversity in nitrate uptake can be observed by reasonably simple assays in a range of agriculturally important plants. However, these simple assays may, or may not, be useful in genotypic screening for purposes of selecting superior types. It is one thing to demonstrate the genetic differences exist; it is something else to identify genotypes which are superior under field conditions. Even so, it appears possible that desirable attributes of the nitrate uptake system can be fostered by suitable genetic manipulations. An understanding of the ways in which the system is regulated may be helpful in this endeavor.

Table 1

RANGE OF NET NITRATE UPTAKE BY THE ROOT
SYSTEMS OF FIVE YOUNG MAIZE INBREDS
DERIVED FROM THE JARVIS POPULATION, AND
THE RANGES IN THE SUBSEQUENT
PARTITIONING OF THE ABSORBED NITRATE TO
REDUCTION, TRANSLOCATION, AND
ACCUMULATION WITHIN THE ROOT TISSUE

	Range	
	Low	High
Mean uptake rate (μmol $g^{-1}_{FW}hr^{-1}$)	4.5 ± 0.16	8.6 ± 0.41
Partitioning (% of uptake)		
Reduction	20.9 ± 5.0	61.6 ± 5.3
Translocation	2.5 ± 1.0	11.4 ± 1.5
Accumulation	30.1 ± 3.0	65.3 ± 5.1

Note: Seedlings were grown in darkness in complete solutions devoid of
nitrogen for 5 to 6 days at which time they were decapitated and
exposed to nitrate-containing solutions 0.35 m*M* for 8 hr. Root
weights ranged from 0.7 to 1.0 g_{FW}. Net nitrate uptake was determined from solution depletion and is presented as the mean of the
constantly increasing rate during the 8 hr. Nitrate translocation
was determined from appearance of nitrate in the bleeding xylem
fluid, and nitrate accumulation by analysis of the tissue. Nitrate
reduction was calculated as the difference between net uptake and
the sum of the quantities translocated and accumulated. From
Pan, W. L., Jackson, W. A., and Moll, R. H., unpublished data,
1981.

III. REGULATION OF THE NITRATE UPTAKE SYSTEM

A. Net Uptake: Influx and Efflux

At concentrations typical of soils solutions, net nitrate uptake by root systems occurs
against a strong electrochemical potential gradient.[74] The electrical component of the
gradient is commonly of the order -100 to -150 mV (inside negative), and the concentration gradient can be more than 100-fold. The resulting large outwardly directed
passive driving force fosters efflux from the tissues. Net efflux to nitrate-free ambient
solutions has been observed when the supply of the carbohydrate to the root systems
was prevented[75] and when high nitrate concentrations had accumulated in the tissue.[60]
Efflux from basal root regions appears greater than from apical regions.[76] Net efflux
also has been observed from roots[53,60,67,77] and from *Lemna paucicostata* fronds[78] to
nitrate-containing solutions. Even when net influx occurs, efflux can still be substantial. In maize, for example, efflux was 20 to 30% of influx during the light period and
higher efflux values were observed in millet.[77] Presence of an exogenous isotopic species of nitrate tends to enhance the measured efflux rate.[60,79] The data can be accommodated by a model[79] in which the ions moving outwardly across the plasmalemma —
through uniporters[80] or through the lipoidal matrix — are visualized as being recycled
within outer unstirred layers to the influx porters or, alternatively, as moving onward
to the ambient solutions. The relative proportion of outwardly moving nitrate that is
reabsorbed depends upon the degree of competition with exogenous nitrate for influx
porter sites and also upon the countering rate of water flow into the root tissue. This
postulate is difficult to test experimentally and it is not possible at the moment to

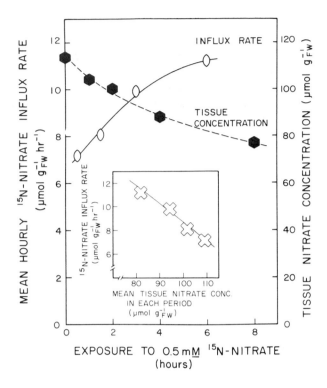

FIGURE 3. Hourly ^{15}N-nitrate influx rates and total root tissue nitrate ([14]N + [15]N) concentrations of five 5-day-old decapitated corn seedlings during exposure to 0.5 mM [15]NO$_3$ (98.7 Atom % [15]N) following an 18-hr pretreatment in 50 mM [14]NO$_3$. The inset shows the relationship between the mean hourly ^{15}N-nitrate influx rates the mean total tissue nitrate ([14]N + [15]N) concentrations during each measurement period.[82]

exclude totally a reciprocal exchange of endogenous and exogenous ions across a common porter with a variable stoichiometry that normally favors the inward movement of the exogenous ions.

Regardless of the precise mechanism responsible for the outward fluxes, differences in net nitrate uptake can be brought about by alterations in influx, efflux, or both. Use of isotopes to estimate both influx and efflux have been employed to examine effects of nitrate preloading,[60,61,81,82] presence of ambient divalent cations[79] and ammonium,[53,62] and diurnal effects.[77] Some divergence in the results is evident. With roots of intact dwarf bean plants, increasing endogenous nitrate concentration accelerated efflux without affecting influx,[53] while in roots of dark-grown maize seedlings, relying on their endosperm as an energy source, influx was restricted temporarily[61] (Figure 3). In roots of intact wheat seedlings, high endogenous nitrate also was associated with decreased influx as well as an accelerated efflux.[60] In this case, however, the restriction in influx persisted for a longer time. Regardless of these differing responses, the general observation of all the isotopic experiments is that the two fluxes may not be directly related; a restriction in influx is not necessarily accompanied by enhanced efflux, nor is the converse necessarily true.

Although the experimental evidence is limited, and the mode of regulation of the two fluxes has not been established, it is clear that knowledge of how net uptake by a plant root system responds to its soil solution and aerial environment ultimately can be attained only by understanding the manner in which both influx and efflux are affected. Such experiments are difficult and time consuming. Moreover, because of the relatively low sensitivity of ^{15}N procedures, experiments employing this isotope commonly have been conducted over time periods which do not permit a precise definition of the absolute influx and efflux rates.[83] If recycling within outer unstirred layers[79] does occur to a significant extent, efflux, as measured by appearance of the endogenous isotopic species to the ambient solution, will underestimate outward movement through the plasmalemma. Similarly, influx as measured by the appearance of the exogenous species in the tissue will underestimate the total nitrate passing through the influx pumps. These potential complexities are exacerbated as the experimental period lengthens because of outward movement of the exogenous ions as their concentration builds up in the root cell cytoplasm. Nevertheless, the magnitude of the values for endogenous nitrate efflux which have been obtained in root tissue[53,60,77] as well as net efflux in *Lemna* fronds,[78] emphasizes that substantial nitrate traffic can occur both ways across the plasmalemma of cells.

Examination of nitrate fluxes under shorter time intervals can be conducted with the use of the short-lived radioisotope ^{13}N[54,84,85] and by using ^{36}Cl-labeled chlorate as an analogue for nitrate.[51,52,86] The limited faculty for proton irradiation of water to produce ^{13}N and the short half-life of the nuclide (10 min) will no doubt limit its use to relatively few laboratories. Chlorate has the potential for more widespread use. A thorough examination with barley[51] revealed similar total uptake at different nitrate-to-chlorate ratios, competitive inhibition in uptake at concentrations to 0.5 mM, similar interactions at higher concentrations, and reasonable half times for efflux from cytoplasm (17 min) and vacuole (20 hr) by standard efflux analysis. At moderate chlorate concentrations, toxic effects that develop after a few hours, presumably from the reduction of chlorate to chlorite by nitrate reductase,[87] limits the usefulness of chlorate to short-term flux studies or as a tracer in nitrate solutions. If differences between chlorate and nitrate occur in the relative reduction rates of the two entering ions, the relative proportion of the two ions which are transferred to the xylem would be estimated incorrectly. In addition, an apparent stimulation in nitrate uptake at low chlorate concentrations,[67] together with the failure to inhibit nitrate uptake in *Pseudomonas fluorescens*,[85] emphasize that chlorate cannot be used indiscriminately; its appropriateness for each experimental tissue should be defined rigorously.[51,52]

Interpretations of much of the current information on regulation of nitrate uptake is limited by lack of knowledge as to how influx and efflux are affected by given treatments. Nevertheless, a number of factors which exert effects on net uptake can be identified, as noted in the following paragraphs. Moreover, the importance of the process in agriculture and ecology, together with the development of methodologies permitting short-term flux measurements, suggest that more definitive statements will be made in the near future about the relative contributions of altered influx and efflux rates to changes in the net uptake rates. In these investigations it must be recognized that plants have a pronounced capability to adapt their uptake rates to nutritional and environmental perturbations.[88]

B. The Driving Force for Nitrate Uptake

A transmembrane acidity gradient (cytoplasm alkaline with respect to the ambient solution) has been proposed to constitute the driving force for nitrate influx into plant tissue across an hydroxyl:nitrate (or bicarbonate:nitrate) antiporter or by cotransport of hydrogen and nitrate ions inwards across a symporter.[89-95] The two possibilities cannot be differentiated experimentally and we shall use the former for descriptive purposes. Generation of alkalinity in the cytoplasm to sustain the driving force is viewed as resulting from three processes: (1) an electrogenic extrusion of protons to the ambient solution via a plasmalemma ATPase; (2) decarboxylation of organic acid an-

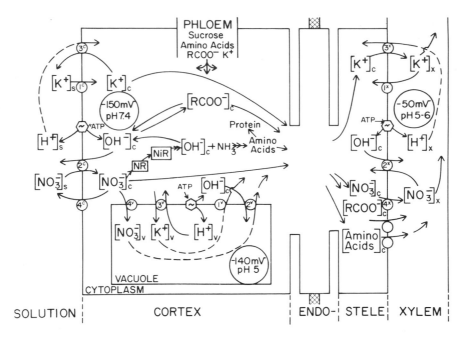

FIGURE 4. Schematic modified from Hanson (1978) for potassium and nitrate transport processes in root tissue and for maintenance of charge balance and intracellular acidity during those processes and during nitrate assimilation in root tissue. Open circles with numbers and arabic superscripts refer to antiporters or uniporters at each membrane; NR, nitrate reductase; NiR, nitrite reductase; RCOO⁻, organic acid anions. Approximate values for acidity in various compartments and for electrical potential gradient between them and the ambient solution are from Raven and Smith.[97]

ions, and (3) nitrate reduction to ammonium, during which one hydroxyl is generated for each nitrate reduced.[96] The salient features are illustrated in Figure 4 which is the trans-root chemiosmotic model of Hanson[80] modified to include the processes involved in nitrate assimilation.[91,92] Approximate values for acidity in the cytoplasm, vacuole, and xylem are from Raven and Smith,[97] and potassium and nitrate are shown as the cation and anion of dominant interest. Presence of ATPase has been reported in xylem parenchyma cells[98] and in vacuole preparations.[99] Net transport across each membrane is viewed as the sum of forward and reverse movement across antiporters and uniporters, with the appropriate driving forces being the acidity and electrical gradients, respectively, across the membranes. Because flux is equal to the driving force divided by the resistance to flow, it can be shown that steady-state deposition in the xylem is possible if the resistance of the porters for each ion differs in the cortical and stelar cell plasmalemmae.[80] Resistance in this context is viewed as including all rate-limiting transport factors including ionic access to and affinity for the porter sites, ionic activities within the various compartments, and number and turnover of the porters.[80]

This concept provides a rationale for maintenance of charge balance in root tissue, ambient medium, and xylem fluid during unequal cation and anion uptake. Consumption of cytoplasmic hydroxyl ions in synthesis of organic acid anions and their efflux during OH^-/NO_3^- antiport provides for maintenance of cytoplasmic acidity within reasonable limits.[100] The concept also envisages the act of nitrate reduction to ammonium as contributing to nitrate influx because of the resulting generation of hydroxyl ions.[96] Positive associations between uptake and reduction[101,102] would be expected provided that the rate of cytoplasmic hydroxyl ion generation by the proton efflux pump was limiting the nitrate influx rate. Increased cytoplasmic alkalinity in the root tissue can also occur indirectly from nitrate reduction in the shoots which fosters synthesis of organic acid anions there, followed by their downward movement in the phloem with the mobile cation potassium and subsequent decarboxylation in the root tissue.[103,104]

A substantial amount of evidence provides general support for the nitrate assimilation and charge balance aspects of the model in higher plants.[92,105] For both tomato[106] and castor bean[107] at various nitrate supplies and harvest times, the sum of organic acid accumulation and hydroxyl ions secreted to the ambient solution was highly correlated with total reduced nitrogen (plus sulfur, which contributes a small proportion of the total cytoplasmic hydroxyl generation), as is required by the model. Moreover, significant downward flow of organic acid anions would be predicted only when (1) nitrate reduction occurs largely in the shoot and, concurrently, (2) net anion uptake significantly exceeds net cation uptake. The former condition apparently obtains in both tomato and castor bean, while the latter occurs in castor bean but not in tomato. In agreement with the prediction, sizable quantities of organic acid anions were detected in phloem sap of the castor bean which, in contrast to tomato, did not respond to increased nitrate supply with increased potassium uptake, perhaps because of enhanced recycling of potassium between shoots and roots.[107]

The relative contribution of downward organic acid anion flow in contributing to the driving force of nitrate influx appears not to be a fixed attribute. At high potassium supplies, recirculation accounted for no more than 20% of the total upward potassium movement in tomato plants, which presumably have relatively little root nitrate reduction, and in which anion uptake and cation uptake were about equal.[108] At low potassium supplies, however, anion uptake exceeded cation uptake substantially, nearly solely due to a decrease in potassium uptake.[109] In spite of much greater uptake of nitrate than potassium, there was only a small decrease in the potassium concentration and no difference in the nitrate/potassium ratio of the xylem fluid. It can be postulated that activity of the H^+-efflux pump was restricted at the low potassium supply owing to the limited cation input, thereby decreasing the cytoplasmic hydroxyl ion generation from this source (Figure 4). Since nitrate uptake was not decreased, cytoplasmic hydroxyl ion generation must have been sustained either by an increase in nitrate reduction within the root tissue[110] or by an increased downward flow (and subsequent decarboxylation) of organic acid anions with potassium. The data thus indicate that sufficient compensation can occur to sustain the driving force for nitrate influx when the H^+-efflux pump activity is restricted. Root nitrate reductase activity appeared not to be altered[109] under the low potassium supply, although this does not exclude the possibility that reduction in the roots did increase. In any case, it would be interesting to know the nature of the signals involved in enhancing nitrate reduction or downward organic acid flow when the potassium supply is limited.

In maize, where ^{15}N experiments indicate significant reduction in roots[58,111-113] but with net anion uptake substantially exceeding net cation uptake, potassium recirculation to the roots appears inadequate to account for an equivalent downward flow and decarboxylation of organic acids sufficient for the excess hydroxyl ion excretion.[114] This view is supported by experiments in which nitrate reduction was inhibited in dark-grown decapitated maize seedlings by endosperm removal; the treatment resulted in decline in the $[NO_3^-/K^+]$ uptake ratio from >1.5 to slightly less than 1.0, largely as a consequence of a decrease in nitrate uptake.[58]

It may also be noted that organic acid anions may contribute to charge balance during upward transport in the xylem when cation uptake exceeds anion uptake,[114-117] when a large proportion of the entering nitrate is reduced in the root system,[58] and when dinitrogen fixation contributes a sizable proportion of the nitrogen input.[93,94,118] Hence, it is evident that there are a number of ways in which metabolism and transport of organic acid anions are intimately involved in nitrate acquisition and assimilation.

The present broad outlines need to be supplemented with more detailed studies of the organic acid anion transport processes.

In spite of the general support for the model proposed in Figure 4, which postulates a direct relationship between the generation of cytoplasmic alkalinity and nitrate uptake, the mechanistic aspects of this association are far from clear. Close correlations between net nitrate uptake and hydroxyl efflux in algae and *Lemna* support the concept of an antiport system for these two ions.[119-121] Inhibiting effects of permeant weak acids[78] which decrease the cytoplasmic hydroxyl ion concentration following their internal dissociation also provide supporting evidence. However, one cannot rule out the possibility of a direct inhibiting effect of cytoplasmic acidity on the nitrate porter system as has been invoked for chloride influx in *Chara*[122,123] and for glucose transport in *Chlorella*.[124,125]

A distinct and characteristic response of the membrane potential to presence of ambient nitrate was observed in energy-rich *Lemna gibba* fronds.[126,127] The typical[128] pattern for H^+-cotransport (or OH^- antiport) was observed: rapid depolarization when substrate nitrate was added followed by a slower repolarization, and transient hyperpolarization occurring when nitrate was removed. The initial depolarization is viewed as a consumption of the proton electrochemical gradient during nitrate transport, while the repolarization in the presence of substrate, and the hyperpolarization following its subsequent removal, are viewed as indicating increased H^+-efflux pump activity as it responds to the decreased membrane potential or increased cytoplasmic acidity. Lower depolarization in light than in dark was attributed to a light-enhanced H^+ efflux pump activity in this tissue.[127,129] The magnitude of the initial depolarization, which suggests initial NO_3^-/OH^- exchange stoichiometry of <1, varied in parallel with measured nitrate uptake rates during induction, and at various ambient concentrations and ambient acidities.[127,129] However, this pattern was observed only with energy-rich frond having membrane potentials more negative than −220 mV. Carbon-starved fronds had much lower membrane potentials (−90 to −120 mV) with little depolarization, or slight hyperpolarization, upon nitrate additions.

In experiments with nitrate-induced excised corn root segments, the membrane potential hyperpolarized upon exposure to nitrate and depolarized upon its removal,[130] which distinctly contrasts with the *Lemna* results. Measurements of net $\bar{H}^+$-efflux to $CaSO_4$ and $Ca(NO_3)_2$ provided evidence for a ratio of 2:1 for net nitrate:hydroxyl antiport. Hyperpolarization still occurred, and depolarization was restricted, upon exposure to nitrate in the presence of 50 mM diethylstilbestrol (DES). Because DES has been shown to inhibit the H^+-efflux ATPase activity, it would appear that the 2:1 ratio was likely not a consequence of an increased H^+-efflux pump activity associated with an actual 1:1 nitrate:hydroxyl antiport. Mitigating somewhat against the 2:1 concept are calculations, assuming reasonable values for cytoplasmic acidity and nitrate concentrations, which indicate the electrochemical gradient for H^+ would be insufficient to drive nitrate uptake at the 2:1 ratio when the external acidity was less than pH 5.5. Thibaud and Grignon[130] point out, however, that high concentration of H^+ in the immediate vicinity of the NO_3^-/OH^- antiporter, as has been suggested to facilitate cotransport in yeast cells,[131,132] could overcome this energetic limitation. They also suggest another possibility; namely, that the induced nitrate uptake system is an anion-hydroxyl ATPase (cf. Reference 133) resulting in electrogenic nitrate uptake.

Together, the membrane potential observations may be considered as indicating intriguing differences between leaf and root tissue in the coupling of nitrate influx and OH^- efflux. Perhaps these differences are as significant as the markedly different regulatory properties of root and shoot nitrate reductase.[134,135] As noted by Ullrich and Novacky,[127] the 2:1 system of the excised roots observed by Thibaud and Grignon could be similar to the low energy system in the *Lemna* fronds. In agreement with this suggestion is that the technique of excision and washing or roots used by Thibaud and Grignon[130] leads to a depletion of carbohydrate stores,[136] although the resting membrane potentials (~-150 mV) they observed doubtless included a sizable energy-dependent component. The possibility of two nitrate uptake systems therefore exists, one associated with a high-energy state characterized by a transient rapid depolarization of the membrane potential, and one associated with a low-energy state involving electrogenic nitrate entry. Two such systems have been identified for phosphate uptake by *Lemna* fronds[137] and two systems for this ion are further implied from the response of excised corn roots to mersalyl.[138]

It may also be noted that the nitrate uptake systems in nitrate-depleted *Chlorella* and *Lemna* cells differ from those of nitrate-grown cells in sensitivity to ambient ammonium, affinity for nitrate, temperature sensitivity, and responsiveness to ambient acidity.[129,139,140] Lower sensitivity of nitrate uptake to ambient ammonium in nitrate-depleted corn roots compared to those in a fully induced state[62,141,142] and a component of the induced system resistant to ammonium in some root tissues[142,143] further support the notion of two nitrate uptake systems which differ at least kinetically. Additional investigations with both root and leaf tissue which include variations in the energy status of each should be helpful in clarifying the nature of the nitrate uptake processes of both. It also will be of interest to determine if at least a part of nitrate uptake is associated with an electrogenic nitrate-stimulated ATPase.[130]

C. The Apparent Induction Pattern

When the plant tissue is first exposed to nitrate there is a relatively low rate of net nitrate uptake which steadily increases to a maximal accelerated rate.[144] In root tissue, a few hours are required for attaining the accelerated rate at ambient concentrations of the order 1 mM or less,[46,47,101,102,141,145-153] although the lag period in *Arabidopsis* was quite short when nitrate uptake was examined at low ambient concentrations.[154]

Only a small amount of nitrate entering the tissue was required to initiate the events leading to the development of the accelerated rate: i.e., less than 5 μmol g_{FW}^{-1} in decapitated maize seedlings.[146] Similarly, exposure to 10 μM nitrate, at which negligible net uptake occurred, was sufficient to overcome the slow initial rate in barley.[50,51] However, the uptake and assimilation of 5.5 μmol $NO_2^- g_{FW}^{-1}$ in corn during a 2-hr pretreatment exposure to $NaNO_2$ failed to hasten the apparent induction phase of nitrate uptake.[146] Nitrate was also shown to be much more effective than nitrite at inducing the nitrate uptake system in roots of dwarf bean.[153] Assuming the products of nitrite assimilation are similar under conditions of nitrite uptake and nitrite production during nitrate reduction, then these observations implicate nitrate itself as the specific inducing agent of the nitrate uptake system in root tissue, rather than the hydroxide ion or reduced nitrogen assimilatory products generated during and sequentially following nitrite reduction.

As noted in the previous section, the induced accelerated system of chlorophyllous tissue differs from the noninduced system in response to a number of experimental conditions.[129,139,140] Root tissue has not been examined as thoroughly for these differential responses. Nevertheless, development of the accelerated rate of net nitrate uptake by root tissue is restricted by inhibitors of ribonucleic acid and protein synthesis,[46,146,148,155,156] suggesting that net synthesis of a component of the uptake system is induced by nitrate. Results of the experiments are not unequivocal, because it is conceivable that a parallel inhibition in induction of nitrate reductase activity could limit nitrate reduction and the associated increase in cytoplasmic alkalinity. The consequence would be a decrease in the driving force for uptake and/or decreased activity of the nitrate:hydroxyl antiporter (Figure 4). Recent experiments, however, indicated that the absolute increase in nitrate reduction by roots of decapitated maize seedling during the induction period was not sufficient to account for the increase in nitrate uptake rates.[102] When considered in conjunction with the nitrite feeding experiments alluded to earlier, these studies have presented evidence which tends to repudiate hypotheses which suggest that hydroxide ion generation during nitrate reduction to ammonia can totally account for the apparent induction phase of nitrate uptake, either via conformational changes in a constitutive transport protein, or by increased driving force. Viewed collectively, the observations amassed thus far favor the proposed mechanism that suggests the apparent induction phase of nitrate uptake in root tissue kinetically characterizes a nitrate-induced synthesis of all or part of a nitrate transport protein.

Various perturbations have been observed in both the shape of the induction pattern and the magnitude of the accelerated rate of nitrate uptake in response to exogenously applied growth regulators.[157] Direct experiments relating environmentally or nutritionally induced changes in endogenous growth regulator balance of the root tissue are required to delineate the potential importance of these effects.

To our knowledge, no definitive evidence is yet available which describes the rate of decay of the nitrate uptake system when root tissues fully induced to the accelerated rate are transferred to nitrate-free media. How the decay patterns are affected by inhibitors of protein synthesis and by the level of end-product effectors should be of value in characterizing the system.

D. Feedback Effects: Transinhibition by Nitrate and Repression or Inhibition by Products of Nitrate Assimilation

When plants are exposed to high supples of any of a number of nutrients their uptake rates for that nutrient frequently become restricted.[64] This has been shown clearly for nitrate in various phytoplankton. For example, the apparent maximal net nitrate uptake rate was much larger than the steady-state rate at relatively low cell nitrogen concentrations.[158] But at high levels of cell nitrogen, the maximal uptake rate declined and approached that of the steady-state rate. Similar patterns can be inferred from the experiments of Deanne-Drummond[50] with barley. Prior exposure to high nitrate supplies, with a corresponding increase in the root tissue nitrate concentration, restricts subsequent net nitrate uptake.[60,63,75,154] With dwarf bean, this effect was entirely accounted for by enhanced nitrate efflux; influx was not affected over a wide range of nitrate pretreatment concentrations[53] (cf. Reference 52). In other experiments, influx apparently was inhibited.[60,61,82]

More rapid recovery of net uptake following transfer to nitrogen-free media was observed in a mutant of *Arabidopsis* with low nitrate reductase activity compared to the wild type.[154] This observation argues against endogenous nitrate exerting transinhibitory effects on net uptake because the prior exposure to nitrate-containing media should have increased root tissue nitrate in the mutant about twofold.[159] On the other hand, parallelism between the decline in total tissue nitrate and the progressive increase in nitrate influx by decapitated maize roots following transfer from high to low ambient nitrate concentrations[61,82] (cf. Figure 3) indicates the possibility of a transinhibitory effect of endogenous (presumably cytoplasmic) nitrate on the influx process (cf. References 90 and 160). It is equally plausible, however, that the initial slow rate of influx and its subsequent recovery resulted from a progressively decreasing concentration of a product of nitrate reduction.

Definitive experiments involving the extent of transinhibition of influx as a regulatory mode in root tissue require methods which unequivocally define root tissue pool sizes for cytoplasmic and vacuolar nitrate and amino acids, the concurrent fluxes into and out of each, and the flux of nitrate through reduction and translocation to the xylem. Presently, the methods available for estimating cytoplasmic nitrate concentrations by modifications of the in vivo nitrate reductase assay remain equivocal,[161-163] although they indicate that it constitutes a relatively small proportion of the total tissue nitrate. Experiments have yielded divergent estimates for the proportion of total nitrate present in vacuoles of leaf tissue although sizable accumulation is indicated.[164,165] Determining nitrate cytoplasmic pool sizes in root systems by compartmental efflux analysis also is not straightforward. Roots contain a number of different types of tissue, the cells of which may differ in the relative sizes of their cytoplasmic pools.[89] Moreover, this approach requires accurate estimates for the concurrent consumption of cytoplasmic nitrate in the reduction and translocation components of the nitrate assimilation pathway during the efflux measurements.

Even it if were possible to define cytoplasmic nitrate concentrations reasonably precisely such that more definitive investigation of transinhibition and efflux as regulatory modes could be undertaken, there is still the strong possibility that repression or inhibition of the nitrate uptake system by products of nitrate assimilation may contribute to the regulation. Prior exposure to ambiently supplied amino acids was found to have pronounced restricting effects on subsequent net nitrate uptake.[154] Interpretation of certain amino acid feeding experiments designed to expose possible feedback inhibition or repression of nitrate assimilation by potential end products of ammonium and/or nitrate assimilation may be complicated by the apparent energy requirement for active transport of amino acids. Depolarization of transmembrane potentials has been observed during amino acid uptake by plant tissues.[126,166-168] Subsequent repolarization in oat coleoptiles was shown to be dependent on an ATP-dependent extrusion of hydrogen ions, and was therefore related to cellular ATP levels.[167] It is conceivable that exogenous application of amino acids may inhibit nitrate transport by competing for the proton motive force required for both uptake processes, irrespective of possible additional allosteric or repression effects on the various components of nitrate assimilation.

We are thus left with a great deal of uncertainty as to the manner in which the nitrogen status of the root tissue can modulate nitrate uptake, although we know that it can do so. A sizable effort is required to delineate the mechanistic aspects of the components of the overall regulation. In pursuing such studies, however, it is necessary to bear in mind that the extent to which negative effectors, be they cytoplasmic nitrate or certain amino acids, exert their restricting effects on influx (and/or in enhancing efflux) will depend on the rate at which they are removed from the region where they exert their inhibitory actions, relative to the rate at which they are deposited there. In contrast to unicellular organisms, roots generally are efficient in translocating solutes out of their cells. The ultimate regulation may well be the rate at which the potential negative effectors are able to be translocated away from locations adjacent to the nitrate influx sites and secreted into the xylem.[110,142]

E. Energy Supply

Results of a series of experiments with controlled environments[55,57,169] are in accord with the concept of an interdependence between net photosynthesis and nitrate absorption.[170,171] When plant growth is maintained under a nonlimiting aerial environment, this concept envisages that shoot growth is largely conditioned by the nitrogen supply, which exerts its primary effect on initiation and expansion of new leaves. Conversely, nitrate absorption during nonlimiting rhizosphere conditions is regulated by the rate of photosynthate translocated to the root system. The amount of photosynthate translocated downward depends upon the amount in excess of that used in growth and respiration of the shoot tissue. Hence, aerial environments which foster high net photosynthesis rates provide sufficient carbohydrate to the root system to sustain high nitrogen uptake rates. Low net photosynthesis results in an insufficient downward transport of carbohydrates, thereby limiting nitrogen uptake rates. Parallelism between the relative growth rates of root systems and the relative nitrogen absorption rates for the whole plants[55-57] under a number of environmental conditions with different plant species support this general concept. It follows that when the external nitrogen supply is limiting, relative to the factors that foster net photosynthesis, the relative growth rate will be a function of the nitrogen supply. On the other hand, variations in the relative growth rate brought about by alterations in other environmental conditions or by genetic constraints will have an impact on the nitrate absorption rate through the energy supply to the root system, with consequent alterations in the nitrate transport processes and/or the surface area available for absorption.

Relatively rapid adverse effects of decapitation or ringing intact plants on nitrate uptake[150,151,172-174] would seem to indicate that the process is strongly dependent upon the energy supply from the shoots. Decreased uptake with decreasing light intensity[72,175-179] the existence of diurnal cycles in nitrate uptake,[43,77,180-182] and higher nitrate uptake rates per unit root dry weight of noninoculated, relative to nodulated, soybean plants[183] also indicate a close dependency. A high carbohydrate requirement for nitrate uptake is further indicated by experiments showing its restriction upon endosperm removal in rice[184] and corn[58] seedlings, and by the requirement of exogenous sugars for sustained nitrate uptake by excised roots.[185,186] Where comparisons are possible, it is clear that nitrate reductase activity or nitrate reduction is more adversely affected than nitrate uptake by treatments resulting in carbohydrate limitations.[58,151,187,188] Maintenance of root nitrate reductase activity is highly dependent upon the carbohydrate status.[185,189,190]

In none of these studies, however, has the total rate of downward flow of carbohydrates from the shoots to the root system been measured directly, Moreover, the effects on nitrate uptake of restricted phloem transport into the root system are not always readily explained as resulting from a decrease in the energy supply. Inhibition of nitrate uptake (but not phosphate uptake) occurred sooner after decapitation than detectable decreases in root respiration[191] or in root tissue glucose and ATP concentrations.[192] Potassium and chloride uptake, and net H^+-efflux, responded similarly to nitrate. Those observations led Bowling[192] to propose a factor(s) of shoot origin which modulates the utilization of energy in the uptake process. Various growth regulators would seem to be logical candidate(s) for the putative controller(s), although no direct evidence is presently available showing that their continual input into the root system is required to sustain nitrate uptake.

There also are other observations which are not totally in accord with direct energy supply limitations. Hänisch Ten Cate and Breteler[151] noted that exogenous sugar additions did not fully restore nitrate uptake rates of excised roots but did restore root nitrate reductase activity. Moreover, they provide intriguing evidence for a specific role of glucose in addition to its influence as an energy source. The data may be in accord with the report[193] that carbon metabolism, in addition to providing the necessary energy, was essential for regulation of nitrate uptake in *Scenedesmus*. Finally, downward malate movement could affect the nitrate uptake process as well as root nitrate reductase activities.[194,105]

Regardless of whether restricted energy supply per se or limited supply of the putative controller(s) is the primary consequence of inhibited phloem transport into root systems, present evidence does not explain whether the effects are exerted on generation of the driving force for nitrate influx, in alterations in activity or net synthesis of the nitrate transport porters, or from initial effects on other component processes of the nitrate assimilation pathway (e.g., translocation) followed by secondary effects restricting influx or accelerating efflux.

F. The Effect of Ammonium

Net nitrate uptake by root systems frequently has been shown to be decreased by the presence of ambient ammonium,[144] although there are instances where it has had relatively little effect.[196-199] Some experiments indicate a residual component of nitrate uptake not sensitive to high ammonium concentrations,[45,46,142,143,150] In other experiments, considerably more sensitivity is indicated; addition of ammonium resulted in an initial net efflux of nitrate, and subsequent net uptake was nearly eliminated at higher ammonium concentrations.[67] On the other hand, ambient ammonium restricted only nitrate influx in dark-grown decapitated corn seedlings; efflux of endogenous nitrate was not enhanced.[62] A specific effect on development of the influx system[141] as well as on its activity[62] seems to be indicated in the corn root tissue.

Because ammonium and certain amino acids restrict nitrate reductase activity in roots,[134,200-202] some of the effects of ambient ammonium in limiting nitrate influx could be ascribed to restriction in the act of nitrate reduction (likely from a product of ammonium assimilation repressing nitrate reductase[203-206] resulting in lowered generation of cytoplasmic alkalinity). However, the decrease in nitrate reduction was not sufficient to account for the total decrease in nitrate influx either under fully induced conditions[62] or during development of the accelerated uptake phase.[141] Under both conditions, ammonium significantly decreased the proportion of the entering nitrate which was reduced. Furthermore, ambient ammonium inhibited nitrate uptake in an *Arabidopsis* mutant deficient in nitrate reductase activity.[154] The evidence clearly suggests direct effects (which may include accelerated efflux[52] as well as restricted influx) on nitrate uptake being exerted by the presence of ambient ammonium. It is not yet clear whether the restrictions result from ammonium interacting with the nitrate porters at the external surface of the plasmalemma or whether they can be attributed to ammonium (or early products of its assimilation) interacting at the internal surface (cf. Reference 207). Indirect evidence supports the notion of internal effects by a product of ammonium assimilation,[142] although this cannot be viewed as conclusive, especially since the inhibition can occur very quickly.[67]

Apart from possible feedback effects by amino acids, pH and energy considerations also may be of importance in regard to the mechanism of inhibition. A rapid depolarization of the membrane potential upon addition of ammonium was observed in *Neurospora*,[208] (cf. Reference 209). Ammonium assimilation into glutamine catalyzed by glutamine synthetase results in net acidification of the cytosol which could result in a decrease in the transmembrane pH gradient. In addition, this first step in ammonium assimilation requires ATP, and may thereby compete for this substrate with plasmalemma ATPases, which as previously discussed may contribute to the driving force of nitrate uptake. Exogenous supplementation of ATP in Nostoc did not fully alleviate ammonium inhibition of nitrate uptake, but it did lessen the severity.[210]

The view that effects are exerted directly on the uptake system and are not totally an indirect consequence of ammonium-induced restrictions in nitrate reduction is also supported by data with other tissue where marked inhibition of nitrate uptake upon addition of ambient ammonium is common. In algae, rapid restrictions occur in nitrate uptake before measurable decreases in nitrate reductase activity are detectable.[205,211-215] Restrictions of nitrate uptake in tungstate-treated cells containing inactive nitrate reductase[216,217] support this view, as do the inhibition kinetics with *Neurospora*[218] and the offsetting of ammonium inhibition by added ATP in *Nostoc*.[210] In none of these studies, however, has the actual rate of reduction been measured. Finally, it again should be noted that when conditions for growth are highly favorable, such that the root system receives a sustained sizable supply of carbohydrates, rapid translocation to the shoots of ammonium or products of its assimilation during periods of high transpiration could militate against the inhibiting effect(s).[110,142] Accordingly, variations in the magnitude of the inhibitory effects of ambient ammonium observed in different experiments with higher plants seem not unreasonable.

IV. NITRATE UPTAKE AND REDUCTION IN ROOT SYSTEMS

The proportions of nitrate and organic nitrogen found in the xylem fluid of decapitated plants differ substantially among species,[114,219-221] and sizable differences have been observed in soybean cultivars.[222] Such data usually are taken to indicate the proportion of the entering nitrate which is reduced in the root system. The method cannot be viewed as exact because the proportion of the organic nitrogen originated from protein turnover in roots or from recycling between roots and shoots is not known[219,223] and specific information about the variation in the magnitude of these processes is limited.[224,225] Even in species where nitrate dominates the xylem, significant reduction in the roots can occur[226] and the process cannot be ignored.

The extent to which nitrate uptake and reduction in root systems are associated processes has apparently varied with the experimental systems examined, and the biochemical mechanisms which underlie these differences have yet to be elucidated. During initial exposure to nitrate, nitrate reductase activity and nitrate uptake rate both increased in roots of dwarf bean,[149] corn,[149,155,227] and rice,[152] as has also been observed in other experimental tissues.[218,228,229] Nevertheless, Breteler and colleagues present substantial evidence showing differential induction patterns for nitrate reductase activity and nitrate uptake rates. The rate of nitrate uptake reached a maximum at 6 hr while in vivo root nitrate reductase activity — which agreed well with $^{15}NO_3^-$ reduction during a 6-hr exposure[150] — continued to increase to 15 hr.[149] Differential effects on the two processes resulted from altered carbohydrates status.[151] Differences in the effects of exogenously supplied growth regulators also occurred, with kinetin stimulating nitrate reductase activity but inhibiting the uptake rate.[157] The response of the two processes to the presence of ambient nitrite differed markedly,[153] as did their response to ambient nitrate concentrations; nitrate reductase activity was constant between 100 μM and 5 mM nitrate whereas the uptake rate exhibited multiphasic patterns.[53]

Nitrate uptake and nitrate reductase activity also can be differentially affected in other tissue. In algae, for example, addition of ammonium resulted in rapid decreases in nitrate uptake before adverse effects were noted in nitrate reductase activity.[205,211-214] Different inhibitory effects of ammonium on nitrate uptake and nitrate reductase activity, and differences in the decay kinetics for the two processes, have been reported in fungi.[218,229] Nitrate uptake increased immediately upon illumination in *Chlorella* whereas nitrate reductase activity required a longer time to respond.[213] In *Lemna* there appears to be a difference during induction in the time course of the increase in nitrate uptake rate[129] and the increase in nitrate reductase activity.[230] Significant uptake and accumulation of nitrate can occur in tissue having inactive nitrate reductase owing to tungstate treatment[217,218,228,231,232] and in mutants defective in nitrate reductase.[47,159,233-238] Moreover, two mutant strains of *Chlamydomonas reinhardi* unable to grow on nitrate had normal nitrate reductase activities implying a separate mutation of the uptake system,[239] although other investigators[240] report that none of their mutations lacking a functional reductase exhibited significant uptake. It therefore appears that substantial uptake can occur when the capacity for reduction is limited, although it has not been established that maximal uptake rates are possible under those conditions.[216,217] Overall, the lack of association between nitrate uptake rates and nitrate reductase activity indicate that nitrate uptake may not be functionally related to nitrate reduction in all instances. This interpretation implies that the two processes may not be structurally coupled as was proposed earlier,[227] although inclusion of a structurally linked uptake and reduction system in a multimechanistic model for nitrate uptake remains possible.

A caveat is necessary, however, because measurements of nitrate reductase activity, either in vivo or in vitro, do not necessarily indicate the actual rate of reduction *in situ*.[144,177,199,234,236,241] Accordingly, some caution must be exerted when interpreting measurements of nitrate reductase activity to indicate the degree of association between uptake and reduction. Even so, *in situ* nitrate reduction in corn roots was affected more adversely than nitrate uptake by the presence of ambient ammonium[62,141,142] and by carbohydrate limitation resulting from endosperm removal.[58,102] In barley seedlings, reduction was more severely restricted by carbohydrate depletion than was uptake.[187] Moreover, an increasing proportion of the entering nitrate being reduced during the induction period[102,156,242] also is counter to a direct association in the two events, as is genetic variation in the proportion of the entering nitrate which is reduced in root systems of maize genotypes (Table 1).

Nevertheless, there are conditions under which a relatively close and persistent association between the two processes can be demonstrated.[101] When dark-grown corn seedlings were exposed to $^{15}NO_3$, a relatively constant proportion of the entering nitrate was reduced when the uptake rate was restricted by prior exposure to high $^{14}NO_3$ concentrations.[61,82] In contrast, concurrent reduction of previously accumulated $^{14}NO_3$ could not be detected under these conditions, although it was translocated to xylem and effluxed to the external solution in sizable quantities. On the other hand, substantial endogenous nitrate was reduced in the absence of exogenous nitrate. The data have been interpreted as indicating that reduction was limited largely to those ions entering the tissue; efflux through the plasmalemma and recycling to influx sites were postulated to be necessary for reduction of the endogenous nitrate.[82] This hypothesis requires that reduction occur during or immediately after influx in this tissue. Recent evidence for localization of nitrate reductase in close association with the plasmalemma in *Neurospora*[243] suggests that the postulate may be plausible. Evidence for membrane-associated nitrate reductase in leaf tissue exists,[244-246] although a specific involvement with the plasmalemma was not demonstrated in those studies. Because endogenous nitrate was accessible for translocation but not for reduction in the presence of exogenous nitrate, MacKown et al.[82] also postulated that endogenous nitrate existed in different pools (which could be separated either longitudinally or horizontally in the root tissue), one of which was largely available for translocation whereas the other was largely available only for efflux and recycling to influx porters with subsequent reduction. The hypothesis advanced from these results differs with the more common concept of ready access of endogenous nitrate to nitrate reductase existing nonsequestered in the bulk cytoplasm. Further investigation is necessary to see whether other root tissues show limited access of endogenous nitrate to reduction sites and concurrent sizable efflux in the presence of exogenous nitrate.

V. NITRATE UPTAKE AND TRANSLOCATION

Translocation of nitrate out of the root system to the shoots is a function of the rate at which it enters the roots minus the rates by which it is diverted to reduction or to accumulation in storage deposits. Nevertheless, substantial evidence indicates that the release of solutes to the xylem is regulated independently of their transport into the root system, and that the regulation is imposed during release from the xylem parenchyma cells[247-249] rather than during radial symplastic transport to them. The evidence derives from morphological attributes of the xylem parenchyma cells[247,250-252] including the apparent presence of ATPases,[98,253] and from the effects of added growth regulators and substances which interfere with normal protein synthesis.[248,249,254-258] The data can be interpreted as indicating that transport from the xylem parenchyma cells to the xylem involves proteins which turn over rapidly and may be under hormonal control, although alternative interpretations are possible.[259]

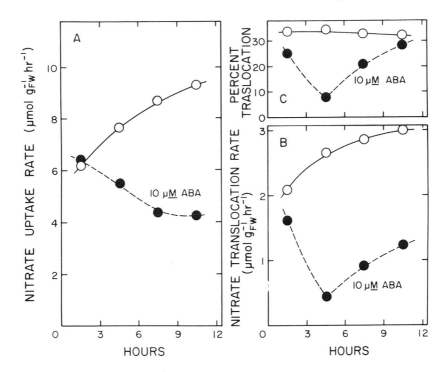

FIGURE 5. Influence of 10 μM abscisic acid (ABA) on mean hourly nitrate uptake (A) and translocation (B) rates, and translocation as a percentage of uptake (C) during four 3-hr periods by dark-grown decapitated maize seedlings exposed to complete solutions containing 0.5 mM KNO$_3$. ABA was added at zero hour. (From Jackson, W. A. and Volk, R. J., unpublished data.)

The investigators who developed this concept did not include nitrate as one of the solutes but the limited data which are available for nitrate release to the xylem indicate the concept is equally applicable for this ion. Nitrate translocation by corn roots is highly sensitive to temperature and anaerobiosis;[148] to 6-methylpurine, an inhibitor of ribonucleic acid synthesis,[144,148,156] to fluorophenylalanine, which results in the synthesis of ineffective protein,[260] and to abscissic acid (Figure 5). In each of these instances, release to the xylem was restricted earlier and to a relatively larger extent than was uptake by the root tissue.

However, complex responses can occur. During continual exposure to ABA, for example, nitrate translocation tended to recover following a severe inhibition whereas there was no recovery in the more modest inhibition of nitrate uptake (Figure 5). Variation in nitrate reduction by the roots[114,219] would influence the amount of nitrate available for translocation.[220,221,261] On the other hand, specific restrictions in nitrate translocation could foster enhanced reduction[221] as has been demonstrated by varying the osmotic water flow through root systems of decapitated corn seedlings.[110] Overall, the data indicate that release of nitrate to the xylem depends strongly on uptake, reduction, and accumulation in storage deposits but that the process may be metabolically regulated such that it is not totally a function of the rate of arrival of nitrate at the xylem deposition sites. The dominant influence, however, is likely to be the extent of reduction in the root system, although that has not yet been established unequivocally. The presence of potassium either externally or internally as a mobile cation facilitates sustained translocation of nitrate.[145,262-265] A cotransport system for the two ions out of the xylem parenchyma cells driven by a potassium concentration gradient, as envisioned for cotransport of potassium and sucrose from mesophyll protoplasts,[266]

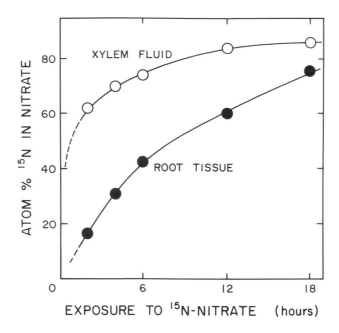

FIGURE 6.   Atom % [15]N in nitrate of xylem fluid and root tissue of 5-day old decapitated corn seedlings during exposure to 0.5 m M (98.25 A % [15]N) following an 18-hr pretreatment in 0.5 m M [14]NO$_3^-$. At the time of transfer the tissue contained 42.3 μmol [14]NO$_3^-$ g$_{FW}^{-1}$.[61]

could explain the results. Alternatively, the data can be accommodated by the trans-root chemiosmotic model[80] which requires transport of a mobile cation from the symplasm to sustain the continual net flux of anions into the xylem (see Figure 4). When K$_2$SO$_4$-treated excised corn roots were transferred to NaNO$_3$, translocation of nitrate was proportional to the electrical potential difference between cortical cells and the xylem sap,[267] lending support to the concepts proposed by Hanson.[80]

Removal of nitrate from the ambient solution results in a relatively rapid decline in its rate of translocation,[81,148,265] although significant quantities, depending on the initial tissue nitrate concentrations, can be translocated at progressively slower rates for a number of hours. When [15]NO$_3^-$ was supplied after prior exposure to [14]NO$_3^-$, the relative proportion in the xylem fluid of exogenous ions to previously accumulated ions exceeded considerably the proportion in the tissue (Figure 6). Entering nitrate clearly was preferentially translocated, compared to endogenous nitrate, even when the root tissue was highly loaded with [14]NO$_3^-$. The data indicate significant compartmentalization of nitrate in root tissue. Nevertheless, ions sequestered in storage pools did have access to translocation channels, albeit less effectively than nitrate entering from the external solution. When nitrate uptake was restricted by the presence of ambient ammonium, translocation of exogenous nitrate was inhibited slightly, but no effect could be detected on translocation of endogenous nitrate.[62] It is likely that the exit of nitrate from storage deposits in root tissue is strongly influenced by the nitrate concentrations in the surrounding cytoplasm. But neither the control mechanisms responsible for the release, not the manner in which nutritional and environmental conditions influence them, have been identified.

As with other ions, nitrate translocation can be enhanced by the flow of water through the root system.[268-270] Accordingly, differences in transpirational water flow may exert effects which mask those observed with decapitated plants which depend primarily on osmotic differences between roots and xylem. Most of the current information on nitrate translocation has been obtained with decapitated root systems because in intact plants it is extremely difficult to ascertain the form in which nitrogen is translocated. The task of determining the extent to which information on nitrate translocation from decapitated plants is directly applicable to intact transpiring plants should be a high priority in future research.

VI. TRANSLOCATION AND THE UPTAKE AND REDUCTION OF NITRATE BY LEAF CELLS

Nitrate reductase in leaf tissue has been investigated intensively and a great deal is known about how its activity is regulated.[271-274] Compared to this wealth of information, knowledge is both uncertain and meager about how the actual quantities of nitrate reduced in leaf cells are regulated. It is difficult to determine with assurance how reductant and substrate supply are regulated in the intact leaf cells. Attempts to do so by measurements of in vivo nitrate reductase activity involving accumulation of nitrite under dark anaerobic conditions have uncertainties resulting from altered membrane permeabilities during the assays such that *in situ* compartmentalization no longer obtains,[163] from observations that nitrite reduction can occur to variable extents under these conditions,[275] and from the possibility of complications due to bacterial contamination as was found in barley root tissue.[276]

Experiments by Shaner and Boyer[270,277] with corn clearly illustrate a strong dependence on the nitrate supplied via the translocation stream in sustaining leaf nitrate reductase activity and a relatively limited ability of high leaf nitrate concentrations to accomplish this purpose. Moreover, some of their data can be interpreted as indicating that reduction was even more dependent on the nitrate translocation rate than was nitrate reductase activity. It was shown that nitrate concentration in xylem exudate collected under pressure immediately after shoot excision, together with previously measured transpiration rates, provided an accurate estimate of the rate of nitrate translocation out of the root tissue and, presumably, the nitrate available for uptake by leaf cells. Chilling the root system,[270] and subjecting plants to low water potentials,[277] decreased nitrate translocation and nitrate reductase activity during periods when leaf nitrate concentrations remained high and did not change significantly. Moreover, varying nitrate movement into excised shoots by altering transpiration (with relative humidity and abscisic acid treatments) or ambient nitrate concentrations showed a correlation between the quantity of nitrate translocated and nitrate reductase activity. In contrast, no relationship was evident between leaf nitrate concentration and leaf nitrate reductase activity. That nitrate reduction was more strongly affected than nitrate reductase activity by the rate of nitrate translocation is evident from the root chilling experiment.[270] Within 3 hr after initial of root chilling, nitrate translocation had decreased to 30% of its initial rate and remained at that rate through 7 hr, whereas nitrate reductase activity only decreased to 70% of the initial rate by the 7th hour. During this time, leaf nitrate concentration did not change, indicating that all the translocated nitrate was reduced. Hence, nitrate reduction was inhibited to the same extent as nitrate translocation; it was restricted sooner and to a greater extent than nitrate reductase activity. These experiments also provide evidence for a slow release of nitrate from storage in corn leaves under natural conditions; calculations revealed movement from vacuoles was more than ten-fold less than the rate of supply from the roots and was less than translocation rates at which leaf nitrate reductase activity decreased appreciably. It is possible, however, that use of stored nitrate for reduction differs among leaf tissue and environmental conditions. Reduction considerably greater than could be accounted for by uptake occurred in 0.5-mm barley leaf slices during darkness.[278]

Although Shaner and Boyer's[270,277] studies leave little question about the important role of the nitrate translocation rate in sustaining leaf nitrate reductase activity, other modes of regulating the synthesis, decay, and activity of the enzyme clearly are important. Thus, Reed and Hageman[279,280] showed that variation in nitrate reductase activity among four maize genotypes could not be related to differences in nitrate translocation. Circadian oscillators and light-on signals, among other factors,[281,282] can contribute to variation in enzyme activity which well may be independent of the nitrate translocation rate. Nevertheless, large variations in nitrate translocation can occur during the daily cycle[226] and its contribution to sustaining leaf nitrate reductase activity should be evaluated before unequivocally invoking other possible reasons for observed differences. The same precaution holds with even greater emphasis for variation in nitrate reduction by leaves, as is indicated by the observation[279,280] that root nitrate uptake and translocation provided a better estimate of reduced nitrogen accumulation among four maize genotypes than did their leaf nitrate reductase activity.

There is little precise information on how nitrate uptake by leaf cells is regulated. Because the relatively small volume of xylem solution which surrounds them has a limited buffering capacity, the leaf cell plasmalemmae must function so as to prevent large differences in net cation and anion fluxes often noted during uptake by root cells. Moreover, the xylem fluid contains an array of organic nitrogen solutes which differs substantially among species and environmental conditions, and which changes as it ascends to leaves of higher position.[219] These substances could have hitherto unrecognized effects on the nitrate uptake process of the leaf cells either directly on the external plasmalemmae surfaces or internally following their absorption.

The development of intercellular compartmentalization of nitrogen assimilatory enzymes in leaf tissue of C_4 species,[283,284] whereby nitrate reductase and nitrite reductase activities are mainly located in mesophyll cells and ammonium-assimilating enzymes are detected in both mesophyll and bundle sheath cells, may well be a developmental or physiological response of the two cell types to the relative proximity of each to the xylem fluid environment. Bundle sheath cells, spatially located between the vascular tissue and mesophyll cells, may serve to buffer against potentially inhibitory effects of highly concentrated, root-derived ammonium and amino acids on nitrate assimilation in the mesophyll. Another ramification of the spatial orientation of these enzymes is the unknown nature of the transport process which links the immediate source of nitrate, the xylem vessel, to the nitrate assimilatory enzymes located in the mesophyll. Mesophyll nitrate concentrations were estimated to be 16 times greater than those found in the bundle sheaths of corn,[284] which suggested the majority of nitrate movement may occur through intercellular spaces between xylem and mesophyll. Alternatively, these concentration differences may only reflect nitrate storage capacity characteristics of each cell type. If nitrate transport does occur intracellularly across the bundle sheath cell en route to the mesophyll cell, the former presents an apparent example of a nitrate uptake system which is not functionally linked to nitrate reduction. If correct, this poses some interesting questions as to the nature of the driving force of nitrate uptake which exists in these cells.

It is important to delineate these and other intriguing aspects (e.g., light dependency), but regulation of nitrate uptake by leaf cells is not an easily examined process. Valuable information can be obtained by placing cut ends of leaf blades (or petioles) in small volumes of solution, monitoring the loss of nitrate from the solution, and measuring the changes in nitrate and nitrite in the leaf tissue.[187] Such experiments provide concurrent estimates of the component processes of the nitrate assimilation pathway. Although nitrate reduction is stimulated by light and carbon dioxide, significant reduction can occur in darkness provided the carbohydrate status of the tissue is not excessively depleted.[187,188,278,285] The data also showed that nitrate reduction was appre ciably more sensitive to energy status than was the estimate of uptake provided by this methodology. It is important to note, however, that the exact quantities of nitrate which enter the leaf cells cannot be determined with certainty by these procedures. Some unknown fraction of the tissue nitrate will be present in the leaf veins. Substantially higher nitrate concentrations occur in midribs of maize[278,279,286] and barley[278] than in the remainder of the lamina, indicating a restriction in transport from midribs into lamina. It is conceivable that high nitrate concentrations may also be accumulating in the minor veins. Since the quantity of nitrate measured as accumulating in the tissue includes that present in veins as well as leaf cells, the nitrate lost from the ambient medium may be an overestimate of net flux across the leaf cell plasmalemmae. Similar reservations apply to ^{15}N experiments; since the sum of [15]nitrate and reduced-^{15}N accumulated in the tissue does not necessarily reflect [15]nitrate entry into the cells.

Lack of quantitative values for nitrate accumulation in veins is a serious methodological problem in determining the extent to which various environmental, nutritional, and genetic factors influence nitrate reduction in leaf cells independently of their effects on nitrate uptake. Correspondingly, it severely complicates assessing the degree of association between the uptake and reduction processes. The problem may be circumvented appreciably by using thin slices of leaf tissue.[278] Use of leaf protoplasts[287] also should provide valuable information, if it is established that none of the processes are perturbed by the isolation procedure.

If the evidence available for nitrate uptake by *Lemna* fronds is a reasonable guide, net nitrate uptake by leaf cells may well be shown to have a complex mode of regulation, including a strong dependency on the intracellular nitrate concentration[78,129] and quite possibly a circadian rhythm.[288] It remains to be seen whether the control systems differ appreciably from those in roots (see Section III.B).

VII. NITROGEN USE EFFICIENCY AND ITS COMPONENT TRAITS

A. Facilitation of Genetic Selection

Genetic materials capable of efficiently utilizing applied nitrogen would be those in which the individual transport and metabolic processes discussed above are well integrated and function most effectively. From the preceding discussion, it is evident that detailed, direct measurement of each of these processes is not possible for evaluating at different levels of nitrogen supply the large numbers of progeny required in a selection program. Even if ways could be found to measure and subsequently improve a small subset of the processes involved, the enormous complexity of the interrelations among the various processes makes it difficult to predict the impact that such improvement would have on overall nitrogen-use efficiency. However, direct assessment of nitrogen-use together with contributions due to its major components is possible from relatively simple measurements of dry matter and nitrogen content in specific plant parts at appropriate times during the growth cycle. These assessments facilitate selection for desirable traits associated with nitrogen acquisition and its utilization in dry matter or grain production, and identify genetic materials most suitable for detailed examination of individual processes. Investigation in detail of these processes in selected genotypes would determine which appear to be most closely associated with efficient traits of nitrogen and carbon assimilation by field crops. The ultimate objective is to facilitate genetic improvement by selection on specific processes.

Moll et al.[12,289] have pointed out a method for using the dry matter and total nitrogen measurements to assess the relative magnitude of some of the components which contribute to efficiency in grain production of maize per unit of nitrogen supplied. These various dry weight and nitrogen values are expressed as a series of ratios, all in units of grams per plant. Thus, nitrogen-use efficiency is defined as (Gw/Ns) where Gw is grain weight and Ns is the nitrogen supplied. Given Nt, the total nitrogen in the plant at maturity, nitrogen-use efficiency can be viewed as the product of two components (Gw/Ns) = (Nt/Ns)(Gw/Nt) where (Nt/Ns) is defined as the nitrogen-uptake efficiency (i.e., the amount absorbed relative to the amount supplied) and (Gw/Nt) is defined as the nitrogen-utilization efficiency (i.e., the effectiveness with which grain is produced per unit of nitrogen absorbed).

In application, this scheme requires measurements of total dry matter, total nitrogen in the plant, and available soil nitrogen. Because these are extremely difficult to measure in field experiments, it is convenient to substitute dry matter and nitrogen in the above-ground tissue and nitrogen supplied as fertilizer, respectively. Possible variation in distribution of nitrogen between roots and shoots, and variation in acquisition of nitrogen from soil sources, remain unevaluated when making these modifications. Caution is therefore necessary when using this approach in making inferences about genetic behavior of plants grown on soil differing greatly in their residual soil nitrogen supply.

Converting the appropriate dry matter and nitrogen ratios to logarithms provides a convenient way for partitioning the variation in (Gw/Ns) among genotypes and at different levels of nitrogen supply to variation in each of its component traits.[12,289] Using this procedure, appreciable differences in (Gw/Ns), (Nt/Ns), and (Gw/Nt) were found among eight single-cross experimental maize hybrids[12] illustrating significant genetic variation in the two traits which contribute to nitrogen-use efficiency. More importantly, the relative contributions of the two traits in this group of hybrids differed between nitrogen fertilizer treatments. At high nitrogen supply, the differences in (Gw/Ns) were largely attributable to variation in nitrogen-uptake efficiency (Nt/Ns), whereas variation in nitrogen-utilization efficiency (Gw/Nt) contributed substantially at low nitrogen supply. The data clearly showed that either high or low values of (Gw/Ns) could be attained by different combinations of the component traits. Equally high (Gw/Ns) values for two hybrids were a result of a relatively high value of (Nt/Ns) in one and a relatively high value of (Gw/Nt) in the other, and different combinations of the two traits also contributed to the low values of (Gw/Ns) in two other hybrids.

Further experiments with two semiprolific hybrids revealed that (Gw/Ns) was substantially larger in two-eared plants than in one-eared plants,[289] with a larger increase in (Nt/Ns) than in (Gw/Nt) of the two-eared plants being responsible. Again, however, differences in the relative contribution of the two components to (Gw/Ns) were observed between levels of nitrogen supply. In this material, a larger advantage in (Nt/Ns) at low nitrogen supply, and a larger advantage in (Gw/Nt) at high nitrogen supply were evident. Other experiments[3,13] confirm the general observation that variation in (Gw/Ns) among genotypes at different levels of nitrogen supply are not a consequence of parallel variation in its component traits.

These findings have important implications for strategies to be used in selection and breeding programs designed to identify materials which are highly efficient in utilization of nitrogen for grain or dry matter production. The soil nitrogen fertility level used in selecting directly for (Gw/Ns), for example, may affect the magnitude and nature of improvement because it would likely affect the selection pressure on each of the two component traits. Moreover, identifying the components of (Gw/Ns) permits the possibility of developing superior genotypes by crossing those with high utilization efficiency with those of low utilization efficiency.[12,289] Finally, the data emphasize that each of the traits involved in nitrogen acquisition is subject to genetic diversity and each may contribute to differing extents under different environmental conditions to the overall effectiveness in nitrogen usage for maximizing grain or dry matter production. The scheme described should therefore aid in the improvement of management practices in general, and should be particularly helpful in the development of specific genotype-management systems to attain maximum nitrogen-use efficiency.

B. Elaboration of the Major Components

Each of the two major components of nitrogen-use efficiency can be subdivided to reflect aspects associated with more specific physiological events. For example, if Tw is the total plant weight at maturity $(Gw/Nt) = (Tw/Nt)(Gw/Tw)$ where Tw/Nt reflects net dry matter produced per unit of N accumulated in the plant, and Gw/Tw reflects the relative distribution of dry matter into grain. Hence, $(Gw/Ns) = (Nt/Ns)(Tw/Nt)(Gw/Tw)$. Alternatively, if Ng represents the nitrogen in the grain at maturity, $Gw/Nt = (Gw/Ng)(Ng/Nt)$ where (Gw/Ng) is the grain weight per unit of nitrogen in the grain and (Ng/Nt) represents the effectiveness with which absorbed nitrogen is translocated to the grain; hence, $(Gw/Ns) = (Nt/Ns)(Gw/Ng)(Ng/Nt)$. Further, for species where a definite transition occurs between vegetative and reproductive growth (e.g., silking in maize), additional insight is afforded by total dry weight and nitrogen measurements at that time. The data provide Nv and Na, the nitrogen absorbed during vegetative and reproductive growth, respectively. Hence, $(Ng/Nt) = (Na/Nt)(Ng/Na)$ where (Na/Nt) is the fraction of total nitrogen absorbed after silking and (Ng/Na) is the ratio of nitrogen translocated to the grain relative to nitrogen taken up during the grain-filling period, which reflects nitrogen remobilized to grain.[12] Hence, nitrogen utilization efficiency can be expressed as $(Gw/Nt) = (Gw/Ng)(Na/Nt)(Ng/Na)$ and nitrogen use efficiency can be expressed as $(Gw/Ns) = (Nt/Ns)(Gw/Ng)(Na/Nt)(Ng/Na)$.

If data on root systems are available, nitrogen uptake efficiency can be subdivided to account for attributes of the root system. For example, (Nt/Ns) can be expanded to reflect the efficiency of N accumulated in roots and the net translocation from roots to tops. Thus, $(Nt/Ns) = (Nb/Ns)(Nt/Nb)$ where Nb is the total N in below-ground plant tissue. Further expansion will take into account effects of N concentration and root mass (Rw). Thus, $(Nb/Ns) = (Rw/Ns)(Nb/Rw)$, and hence $(Nt/Ns) = (Rw/Ns)(Nb/Rw)(Nt/Nb)$.

These kinds of expanded expressions of nitrogen-use efficiency provide bases for more detailed investigation of effects associated with the major component traits, and thereby focus attention on important issues relevant to an understanding of nitrogen-use efficiency. For example, in one study of variation among maize hybrids, the contribution of utilization efficiency (Gw/Nt) to variation in nitrogen-use efficiency (Gw/Ns) was found to be much larger at low nitrogen supply than at high N supply.[12] More detailed analysis revealed that this difference could be largely attributed to differences contributed by proportions of nitrogen translocated to grain (Ng/Nt) and by differences in redistribution of stored N to grain as reflected by (Ng/Na). Furthermore, the greater contributions of both traits, (Ng/Nt) and (Ng/Na), at low nitrogen resulted from higher correlations with nitrogen-use efficiency rather than greater variation in the traits.[12] This means that differences among those hybrids in the proportion of absorbed nitrogen that was translocated to grain were more important in contributing to high N-use efficiency when nitrogen was limiting than when nitrogen was abundant. This, in turn, appears to have been fostered by differences among the hybrids in their ability to redistribute that portion of nitrogen taken up prior to silking. However, the data clearly demonstrate that the magnitude of variation by itself is not a reliable indicator of the importance of a particular component process or trait; its degree of correlation with nitrogen-use efficiency must also be considered.

ACKNOWLEDGMENTS

Paper No. 8735 of the Journal Series of the North Carolina Agricultural Research Service, Raleigh, N.C. We thank H. Breteler for kindly sending us manuscripts prior to their publication, and our colleagues, R. J. Volk, D. W. Israel, C. D. Raper, Jr., and R. C. Fites for their helpful suggestions.

REFERENCES

1. Smith, N., Response of inbred lines and crosses in maize to variations of nitrogen and phosphorus supplied as nutrients, *J. Am. Soc. Agron.,* 26, 785, 1934.
2. Springfield, G. H. and Salter, R. M., Differential response of corn varieties to fertility levels and to seasons, *J. Agric. Res.,* 49, 991, 1934.
3. Kamprath, E. J., Moll, R. H., and Rodriguez, N., Effects of nitrogen fertilization and recurrent selection on performance of hybrid populations of corn, *Agron. J.,* 74, 955, 1982.
4. Hay, R. E., Early, E. B., and Deturk, E. E., Concentration and translocation of nitrogen compounds in the corn plant *(Zea mays)* during grain development, *Plant Physiol.,* 28, 606, 1953.
5. Hanway, J. J., Corn growth and composition in relation to soil fertility. II. Uptake of N, P, and K and their distribution in plant parts during the growing season, *Agron. J.,* 54, 222, 1962.
6. Jung, P. E., Peterson, L. A., and Schrader, L. E., Response of irrigated corn to time, rate and source of applied N on sandy soils, *Agron. J.,* 64, 668, 1972.
7. Beauchamp, E. G., Kannenberg, L. W., and Hunter, R. B., Nitrogen accumulation and translocation in corn genotypes following silking, *Agron. J.,* 68, 418, 1976.
8. Chevalier, P. and Schrader, L. E., Genotypic differences in nitrate absorption and partitioning of N among plant parts in maize, *Crop Sci.,* 17, 897, 1977.
9. Moll, R. H. and Kamprath, E. J., Effects of population density upon agronomic traits associated with genetic increases in yield of *Zea mays* L., *Agron. J.,* 69, 81, 1977.
10. Pollmer, W. G., Eberhard, D., Klein, D., and Dhillon, B. S., Genetic crontrol of nitrogen uptake and translocation in maize, *Crop Sci.,* 19, 82, 1979.
11. Muruli, B. I. and Paulsen, G. M., Improvement of nitrogen use efficiency and its relationship to other traits in maize, *Maydica,* 26, 63, 1981.
12. Moll, R. H., Kamprath, E. J., and Jackson, W. A., Analysis and interpretation of factors which contribute to efficiency of nitrogen utilization, *Agron. J.,* 74, 562, 1982.
13. Anderson, E. L., Kamprath, E. J., and Moll, R. H., Effect of prolificacy and N rate on grain yield, dry matter and N accumulation and partitioning in maize, *Agron. J.,* 74, 1982.
14. Goodman, P. J., Selection for nitrogen responses in Lolium, *Ann. Bot. (London),* 41, 243, 1977.
15. Goodman, P. J., Genetic control of inorganic nitrogen assimilation of crop plants, in *Nitrogen Assimilation of Plants,* Hewitt, E. J. and Cutting, C. V., Eds., Academic Press, 1979.
16. Alagarswamy, G. and Bidinger, F. R., Nitrogen uptake and utilization by pearl millet [Pennisetum americanum (2) Leeke], in *9th Int. Plant Nutrition Colloquium,* Vol. 1, Scaife, A., Ed., Commonwealth Agricultural Bureaux, Farnham Royal, 1982, 12.
17. Haynes, R. J. and Goh, K. M., Ammonium and nitrate nutrition of plants, *Biol. Rev.,* 53, 465, 1978.
18. Clarke, A. L. and Barley, K. P., The uptake of nitrogen from soils in relation to solute diffusion, *Austr. J. Soil Res.,* 6, 75, 1968.
19. Schrader, L. E. and Thomas, R. J., Nitrate uptake, reduction and transport in the whole plant, in *Nitrogen and Carbon Metabolism,* Bewley, J. D., Ed., Martinus Nijhoff/Dr. W. Junk, The Hague, 1982, 49.
20. Spencer, T. A., A comparative study of the seasonal development of some inbred lines and hybrids of maize, *J. Agric. Res.,* 61, 521, 1940.
21. Nass, H. G. and Zuber, M. S., Correlation of corn *(Zea mays* L.) roots early in development to mature root development, *Crop Sci.,* 11, 655, 1971.
22. Raper, C. D. and Barber, S. A., Rooting systems of soybeans. I. Differences in root morphology among varieties, *Agron. J.,* 62, 581, 1970.
23. Sullivan, J. P. and Brun, W. A., Effect of root genotype on shoot water relations in soybeans, *Crop Sci.,* 15, 319, 1975.
24. Schenk, M. K. and Barber, S. A., Root characteristics of corn genotypes as related to P uptake, *Agron. J.,* 71, 921, 1979.

25. Kiesselbach, T. A. and Weihing, R. H., The comparative root development of selfed lines of corn and their F_1 and F_2 hybrids, *Agron. J.,* 27, 538, 1935.
26. Hurd, E. A., Phenotype and drought tolerance in wheat, *Agric. Meterol.,* 14, 39, 1974.
27. Zobel, R. W., The genetics of root development, in The *Development and Functions of Roots,* Torrey, J. G. and Clarkson, D. T., Eds., Academic Press, New York, 1975, 261.
28. Jordan, W. R., Miller, F. R., and Morris, D. E., Genetic variation in root and shoot growth of sorghum in hydroponics, *Crop Sci.,* 19, 468, 1979.
29. Jordan, W. R., Sorghum root growth and morphology, and field rooting systems, in *Plant Roots,* Frey, K. J., Ed., Agronogy Dept. Publication, Iowa State University, Ames, Iowa, 1980, 129.
30. Jordan, W. R. and Miller, F. R., Genetic variability in sorghum root systems: implications from drought tolerance, in *Adaptation of Plants to Water and High Temperature Stress,* Turner, N. C. and Kramer, P. J., Eds., John Wiley & Sons, New York, 1980, 383.
31. Russell, R. S., *Plant Root Systems: Their Function and Interaction with the Soil,* McGraw Hill, New York, 1977.
32. Hackett, C., A method of applying nutrients locally to roots under controlled conditions, and some morphological effects of locally applied nitrate on the branching of wheat roots, *Austr. J. Biol. Sci.,* 25, 1169, 1972.
33. Drew, M. C., Saker, L. R., and Ashley, T. W., Nutrient supply and the growth of the seminal root system in barley. I. The effect of nitrate concentration on the growth of axes and laterals, *J. Exp. Bot.,* 24, 1189, 1973.
34. Drew, M. C. and Saker, L. R., Nutrient supply and the growth of the seminal root system in barley. II. Localized compensatory increases in lateral root growth and rates of nitrate uptake when nitrate supply is restricted to only part of the root system, *J. Exp. Bot.,* 26, 79, 1975.
35. Tennant, D., Root growth of wheat. I. Early patterns of multiplication and extension of wheat roots including effects of levels of nitrogen, phosphorus and potassium, *Austr. J. Agric. Res.,* 27, 183, 1976.
36. Maizlish, N. A., Fritton, and Kendall, W. A., Root morphology and early development of maize at varying levels of nitrogen, *Agron. J.,* 72, 25, 1980.
37. Ingestad, T. and Lund, A. B., Nitrogen stress in birch seedlings. I. Growth techniques and growth, *Physiol. Plant.,* 45, 137, 1979.
38. Ingestad, T., Nitrogen stress in birch seedlings. II. N, K, P, Ca and Mg nutrition, *Plant Physiol.,* 45, 149, 1979.
39. Ingestad, T., Growth, nutrition and nitrogen fixation in grey alder at varied rate of nitrogen addition., *Physiol. Plant.,* 50, 353, 1980.
40. Ericsson, T., Effects of varied nitrogen stress on growth and nutrition in three Salix clones, *Physiol. Plant.,* 51, 423, 1981.
41. Ericsson, T., Larsson, C.-M., and Tillberg, E., Growth responses of Lemna to different levels of nitrogen limitation, *Z. Pflanzenphysiol.,* 105, 331, 1982.
42. Edwards, D. G. and Asher, C. J., The significance of solution flow rate in flowing culture experiments, *Plant Soil,* 41, 161, 1974.
43. Clement, C. R., Hopper, M. J., Jones, L. H. P., and Leafe, E. L., The uptake of nitrate by Lolium perenne from flowing nutrient solution. II. Effect of light, defoliation, and relationship to CO_2 flux, *J. Exp. Bot.,* 29, 1173, 1978.
44. van den Honert, T. H. and Hooymans, J. J. M., On the absorption of nitrate by maize in water culture, *Acta Bot. Neerl.,* 4, 376, 1955.
45. Lycklama, J. C., The absorption of ammonium and nitrate by perennial ryegrass., *Acta Bot. Neerl.,* 12, 316, 1963.
46. Rao, P. K. and Rains, D. W., Nitrate abosprtion by barley. I. Kinetics and energetics, *Plant Physiol.,* 57, 55, 1976.
47. Doddema, H., Hofstra, J. J., and Feenstra, W. J., Uptake of nitrate by mutants of *Arabidopsis thaliana,* disturbed in uptake or reduction of nitrate. II. Effect of nitrogen source during growth on uptake of nitrate and chlorate, *Physiol. Plant.,* 43, 343, 1978.
48. Bhat, K. K. S., Nye, P. H., and Brereton, A. J., The possibility of predicting solute uptake and plant growth response from independently measured soil and plant characteristics. VI. The growth and uptake of rape in solutions of constant nitrate concentration, *Plant Soil,* 53, 137, 1979.
49. Van de Dijk, S. J., Lanting, L., Lambers, H., Posthumus, F., Stulen, I., and Hofstra, R., Kinetics of nitrate uptake by different species from nutrient-rich and nutrient-poor habitats as affected by the nutrient supply, *Physiol. Plant.,* 55, 103, 1982.
50. Deanne-Drummond, C. E., Mechanisms for nitrate uptake into barley (*Hordeum vulgare,* cv. Fergus) seedlings grown at controlled nitrate concentrations in the nutrient medium, *Plant Sci. Lett.,* 24, 79, 1982.
51. Deanne-Drummond, C. E. and Glass, A. D. M., Nitrate uptake into barley (*Hordeum vulgare* plants: a new approach using $^{36}ClO_3^-$ as an analog for NO_3^-, *Plant Physiol.,* 70, 50, 1982.

52. Deanne-Drummond, C. E. and Glass, A. D. M., Short term studies of nitrate uptake into barley plants using ion specific electrodes and $^{36}ClO_3^-$. I, II, 73, 100, 105, 1983.

53. Breteler, H. and Nissen, P., Effect of exogenous and endogenous nitrate concentration on nitrate utilization by dwarf bean, *Plant Physiol.,* 70, 754, 1982.

54. Thayer, J. R. and Huffaker, R. C., Kinetic evaluation, using ^{13}N, reveals two assimilatory nitrate transport systems in *Klebsiella pneumoniae, J. Bacteriol.,* 149, 198, 1982.

55. Raper, C. D., Jr., Parsons, L. R., Patterson, D. R., and Kramer, P. J., Relationship between growth and nitrogen accumulation for vegetative cotton and soybean plants, *Bot. Gaz. (Chicago),* 138, 129, 1977.

56. Raper, C. D., Jr., Relative growth and nutrient accumulation rates for tobacco, *Plant Soil,* 46, 473, 1977.

57. Raper, C. D., Jr., Osmond, D. L., Wann, M., and Weeks, W. W., Interdependence of root and shoot activities in determining nitrogen uptake rates of roots, *Bot. Gaz. (Chicago),* 139, 289, 1978.

58. Jackson, W. A., Volk, R. J., and Israel, D. W., Energy supply and nitrate assimilation in root systems, in *Carbon-Nitrate Interaction in Crop Production,* Tanaka, A., Ed., Japan Society for the Promotion of Science, Tokyo, 1980, 25.

59. Edwards, J. H. and Barber, S. A., Nitrogen flux into corn roots as influenced by shoot requirement, *Agron. J.,* 68, 471, 1976.

60. Jackson, W. A., Kwik, K. D., Volk, R. J., and Butz, R. G., Nitrate influx and efflux by intact wheat seedlings: effects of prior nitrate nutrition, *Planta,* 132, 149, 1976.

61. MacKown, C. T., Volk, R. J., and Jackson, W. A., Nitrate accumulation, assimilation, and transport by decapitated corn roots, *Plant Physiol.,* 68, 133, 1981.

62. MacKown, C. T., Jackson, W. A., and Volk, R. J., Restricted nitrate influx and reduction in corn seedlings exposed to ammonium, *Plant Physiol.,* 69, 353, 1982.

63. Clement, C. R., Jones, L. H. P., and Hopper, M. J., Uptake of nitrogen from flowing nutrient solution: effect of terminated and intermittent nitrate supplies, in *Nitrogen Assimilation of Plants,* Hewitt, E. J. and Cutting, C. V., Eds., Academic Press, New York, 1979, chap. II.2.

64. Clarkson, D. T. and Hanson, J. B., The mineral nutrition of higher plants, *Annu. Rev. Plant Physiol.,* 31, 239, 1980.

65. Bloom, A. J. and Chapin, F. S., Differences in steady-state net ammonium and nitrate influx by cold- and warm-adapted barley varieties, *Plant Physiol.,* 68, 1064, 1981.

66. Rao, P. K., Rains, D. W., Qualset, C. O., and Huffaker, R. C., Nitrogen nutrition and grain protein in two spring wheat genotyupes differing in nitrate reductase activity, *Crop Sci.,* 17, 238, 1977.

67. Doddema, H. and Telkamp, G. P., Uptake of nitrate by mutants of *Arabidopsis thaliana,* disturbed in uptake or reduction of nitrate. II. Kinetics, *Physiol. Plant.,* 45, 332, 1979.

68. Jones, P. W., Johnson, C. B., and Whittington, W. J., Nitrate uptake and the induction of nitrate reductase activity in wheat (*Triticum aestivum* L.,) seedlings, *J. Exp. Bot.,* 32, 9, 1981.

69. Misra, B. B., Behera, P. K., and Tripathey, P. C., Comparative studies on nitrate uptake and nitrate reduction in root and shoot of rice seedlings (*Oryza sativa* L.) induced by nitrate, *Plant Biochem. J.,* 8, 97, 1981.

70. Baer, G. R. and Collet, G. F., *In vivo* determination of parameters of nitrate utilization in wheat (*Triticum aestivum* L.) seedlings grown with low concentration of nitrate in the nutrient solution, *Plant Physiol.,* 68, 1237, 1981.

71. Polisetty, R. and Hageman, R H., Studies on nitrate uptake by solution grown corn (*Zea mays* L.) genotypes, *Biol. Plant,* 24, 117, 1982.

72. Ta, T. C. and Ohira, K., Comparison of the uptake and assimilation of ammonium and nitrate in indica and japonica rice plants using the tracer ^{15}N method, *Soil Sci. Plant Nutr.,* 28, 79, 1982.

73. Mugwira, L. M., Elgawhary, S. M., and Allen, A. E., Nitrate uptake effectiveness of different cultivars of Triticale, wheat, and rye, *Agron. J.,* 72, 585, 1980.

74. Higinbotham, N., Electropotentials of plant cells, *Annu. Rev. Plant Physiol.,* 24, 25, 1973.

75. Minotti, P. L. and Jackson, W. A., Nitrate reduction in the roots and shoots of wheat seedlings, *Planta,* 95, 36, 1970.

76. Shone, M. T. G. and Wood, A. V., Longitudinal movement and loss of nutrients, pesticides, and water in barley roots, *J. Exp. Bot.,* 28, 872, 1977.

77. Pearson, C. J., Volk, R. J., and Jackson, W. A., Daily changes in nitrate influx, efflux and metabolism in maize and pearl millet, *Planta,* 152, 319, 1981.

78. Löppert, H. and Kronberger, W., Control of nitrate uptake by photosynthesis in *Lemna paucicostata* 6746, in *Photosynthesis Plant Development Proceedings Conference,* Marcelle, R., Clijsters, H., and Van Poucke, M., Eds., W. Junk, The Hague, 1979.

79. Morgan, M. A., Volk, R. J., and Jackson, W. A., Simultaneous influx and efflux of nitrate during uptake by perennial ryegrass, *Plant Physiol.,* 51, 267, 1973.

80. Hanson, J. B., Application of the chemiosmotic hypothesis to ion transport across the root, *Plant Physiol.,* 62, 402, 1978.

81. Jackson, W. A. and Volk, R. J., Nitrate transport processes and compartmentation in roots systems, in *Genetic Engineering of Symbiotic Nitrogen Fixation and Conservation of Fixed Nitrogen,* Lyons, J. M., Valintine, R. C., Phillips, D. A., Rains, D. W., and Huffaker, R. C., Eds., Plenum Press, New York, 1981, 517.

82. MacKown, C. T., Jackson, W. A., and Volk, R. J., Partitioning of previously-accumulated nitrate to translocation reduction, and efflux in corn roots, *Planta,* 157, 8, 1983.

83. Cram, W. J., Influx isotherms — their interpretation and use, in *Membrane Transport in Plants,* Zimmermann, U. and Dainty, J., Eds., Springer-Verlag, Basel, 1974, 334.

84. Skokut, T. A., Wolk, C. P., Thomas, J., Meeks, J. C., Shaffer, P. W., and Chien, W. S., Initial organic products of assimilation of [^{13}N] ammonium and [^{13}N] nitrate by tobacco cells cultured on different sources of nitrogen, *Plant Physiol.,* 62, 299, 1978.

85. Betlach, M. R., Tiedje, J. M., and Firestone, R. B., Assimilatory nitrate uptake in *Pseudomonas fluorescens* studied using nitrogen-13, *Arch. Microbiol.,* 129, 135, 1981.

86. Tromballa, H. W. and Broda, E., Das verhalten von *Chlorella fusca* gegenüber Perchlorat und Chlorat, *Arch. Mikrobiol.,* 78, 214, 1971.

87. Hofstra, J. J., Chlorate toxicity and nitrate reductase activity in tomato plants, *Physiol. Plant.,* 41, 65, 1977.

88. Rufty, T. W., Jr., Raper, C. D., Jr., and Jackson, W. A., Nitrate uptake, root and shoot growth, and ion balance of soybean plants during acclimation to root-zone acidity, *Bot. Gaz. (Chicago),* 143, 5, 1982.

89. Hodges, T. K., Ion absorption by plant roots, *Adv. Agron.,* 25, 163, 1973.

90. Smith, F. A., The internal control of nitrate uptake into excised barley roots with differing salt contents, *New Phytol.,* 72, 769, 1973.

91. Raven, J. A. and Smith, F. A., Nitrogen assimilation and transport in vascular land plants in relation to intracellular pH regulation, *New Phytol.,* 76, 415, 1976.

92. Kirkby, E. A., Plant growth in relation to nitrogen supply, in *Terrestrial Nitrogen Cycles,* Clark, F. E. and Rosswall, T., Eds., Swedish National Research Council, *Ecol. Bull. (Stockholm),* 33, 249, 1981.

93. Israel, D. W. and Jackson, W. A., The influence of nitrogen nutrition on ion uptake and translocation by leguminous plants, in *Mineral Nutrition of Legumes in Tropical and Subtropical Soils,* Andrew, C. S. and Kamprath, E. J., Eds., Commonwealth Scientific and Industrial Research Organization, Melbourne, Australia, 1978, 113.

94. Israel, D. W. and Jackson, W. A., Ion balance, uptake, and transport process in N_2-fixing and nitrate- and urea-dependent soybean plants, *Plant Physiol.,* 69, 171, 1982.

95. Dejaegere, R., Neirinckx, L., Stassart, J. M., and Delegher, V., Mechanism of ion uptake across barley roots, in *Structure and Function of Plant Roots,* Brouwer, R., Ed., Junk Publ., The Hague, 1981, 173.

96. Dijkshoorn, W., Metabolic regulation of the alkaline effect of nitrate utilization in plants, *Nature (London),* 194, 165, 1962.

97. Raven, J. A. and Smith, F. A., Characteristics, functions, and regulation of active proton extrusion, in *Regulation of Cell Membrane Activities in Plants,* Marrè, E. and Ciferri, O., Eds., North-Holland, Amsterdam, 1977, 25.

98. Winter-Sluiter, E., Lauchli, A., and Kramer, D., Cytochemical localization of K+-stimulated adenosine triphosphatose activity in xylem parenchyma cells of barley roots, *Plant Physiol.,* 60, 923, 1977.

99. Wagner, G. J. and Lin, W., An active proton pump of intact vacuoles isolated from *Tulipa* petals, *Biochim. Biophys. Acta,* 689, 261, 1982.

100. Smith, F. A. and Raven, J. A., Intracellular pH and its regulation, *Annu. Rev. Plant Physiol.,* 30, 289, 1979.

101. Chantarotwong, W., Huffaker, R. C., Miller, B. L., and Granstedt, R. C., *In vivo* nitrate reduction in relation to nitrate uptake, nitrate content, and *in vitro* nitrate reductase activity in intact barley seedlings, *Plant Physiol.,* 57, 519, 1976.

102. Morgan, M. A., Volk, R. J., and Jackson, W. A., Uptake and assimilation of nitrate by corn roots during and after induction of the nitrate uptake system, *J. Exp. Bot.,* 36, 859, 1985.

103. Dijkshoorn, W., Lathwell, D. J., and DeWit, C. T., Temporal changes in carboxylate content of ryegrass with stepwise changes in nutrition, *Plant Soil,* 29, 369, 1968.

104. Ben-Zioni, A., Voadia, Y., and Lips, S. H., Nitrate uptake by roots as regulated by nitrate reduction products of the shoots, *Physiol. Plant.,* 24, 288, 1971.

105. Kirkby, E. A., Recycling in plants considered in relation to ion uptake and organic acid accumulation, in Plant Analysis and Fertilizer Problems, Proc. 7th Int. Colloquium, Hanover, Federal Republic of Germany, 1974, 557.

106. Kirkby, E. A. and Knight, A. H., Influence of the level of nitrate nutrition on ion uptake and assimilation, organic acid accumulation, and cation-anion balance in whole tomato plants, *Plant Physiol.,* 60, 349, 1977.

107. Kirkby, E. A. and Armstrong, M. J., Nitrate uptake by roots as regulated by nitrate assimilation in the shoot of castor oil plants, *Plant Physiol.,* 65, 286, 1980.

108. Armstrong, M. J. and Kirkby, E. A., Estimation of potassium recirculation in tomato plants by comparison of the rates of potassium and calcium accumulation in the tops with their fluxes in the xylem stream, *Plant Physiol.,* 63, 1143, 1979.

109. Kirkby, E. A., Armstrong, M. J., and Leggett, J. E., Potassium recirculation in tomato plants in relation to potassium supply, *J. Plant Nutr.,* 3, 955, 1981.

110. Rufty, T. W., Jr., Jackson, W. A., and Raper, C. D., Jr., Nitrate reduction in roots as affected by the presence of potassium and by flux of nitrate through the roots, *Plant Physiol.,* 68, 605, 1981.

111. Ivanko, S. and Ingversen, J., Investigation on the assimilation of nitrogen by maize roots and the transport of some major nitrogen compounds by xylem sap. III. Transport of nitrogen compounds by xylem sap, *Physiol. Plant.,* 24, 355, 1971.

112. Ivanko, S. and Maxianova, A., Nitrogen uptake and assimilation by maize roots, in *Proc. Int. Symp. Structure and Function of Primary Root Tissues,* Kolek, J., Ed., Veda, Slovak Academy of Sciences, Bratislava, Czechoslovakia, 1974, 461.

113. Yoneyama, T., Akiyama, Y., and Kumazawa, K., Nitrogen uptake and assimilation by corn roots, *Soil Sci. Plant Nutr.,* 23, 85, 1977.

114. Keltjens, W. G., Nitrogen metabolism and K-recirculation in plants, in *Proc. 9th Annu. Plant Nutrition Colloquium,* Scaife, A., Ed., Commonwealth Agricultural Bureaux, Warwick University, England, 1982, 1, 283.

115. Dijkshoorn, W., Organic acids and their role in ion uptake, in *Chemistry and Biochemistry of Herbage,* Vol. 2, Butler, G. W. and Bailey, R. W., Eds., Academic Press, New York, 1973, 163.

116. Triplett, E. W., Barnett, N. M., and Blevins, D. G., Organic acids and ionic balance in xylem exudate of wheat during nitrate or sulfate absorption, *Plant Physiol.,* 65, 610, 1980.

117. Butz, R. G. and Long, R. C., L-malate as an essential component of the xylem fluid of corn seedlings roots, *Plant Physiol.,* 64, 684, 1979.

118. Van Beusichem, M. L., Nutrient absorption of pea plants during dinitorgen fixation. I. Comparison with nitrate nutrition, *Neth. J. Agric. Sci.,* 29, 259, 1981.

119. Loppert, H. G., Kronberger, W., and Kandeber, P., Correlation between nitrate uptake and alkalinization by *Lemna paucicostata* 6746, in *Transmembrane Ionic Exchanges in Plants,* Thellier, M., Monnier, A., Demarty, M., and Dainty, J., Eds., Colloque International du Centre National del la Researche Scientific No. 258, Paris, 1977, 283.

120. Ullrich, W. R. and Eisele, R., Relations between nitrate uptake and reduction *Ankistrodesmus braunii,* in *Transmembrane Ionic Exchange in Plants,* Thellier, M., Monnier, A., Demarty, M., and Dainty, J., Eds., Colloque International du Centre National de la Researche Scientific No. 258, Paris, 1977, 307.

121. Eisele, R. and Ullrich, W. R., Effect of glucose and CO_2 on nitrate uptake and coupled OH^- flux in *Ankistrodesmus braunii, Plant Physiol.,* 59, 18, 1977.

122. Sanders, D., The mechanism of Cl^- transport at the plasma membrane of *Chora corallina.* I. Cotransport with H^+ *J. Membr. Biol.,* 53, 129, 1980.

123. Sanders, D. and Hansen, U-P., Mechanisms of Cl^- transport at the plasma membrane of *Chara corallina.* II. Transinhibition and the determination of H^+/Cl^- binding order from a reaction kinetic model, *J. Membr. Biol.,* 8, 139, 1981.

124. Komar, E., Schwab, W. G. W., and Tanner, W., The effect of intracellular pH on the rate of hexose uptake in *Chlorella, Biochim. Biophys. Acta,* 55, 524, 1979.

125. Tanner, W., Proton sugar co-transport in lower and higher plants, *Ber. Dtsch. Bot. Ges.,* 93, 167, 1980.

126. Novacky, A., Fischer, E., Ullrich-Everius, C. I., Luttge, U., and Ullrich, W. R., Membrane potential changes during transport of glycine as a neutral amino acid and nitrate in *Lemna gibba* G1, *FEBS Lett.,* 88, 264, 1978.

127. Ullrich, W. R. and Novacky, A., Nitrate-dependent membrane potential changes and their induction in *Lemna gibba* G1, *Plant Sci. Lett.,* 22, 211, 1981.

128. Jung, K.-D., Luttge, U., and Fischer, E., Uptake of neutral and acidic amino acids by *Lemna gibba* correlated with the H^+-electrochemical gradient at the plasmalemma, *Physiol. Plant.,* 55, 351, 1982.

129. Ullrich, W. R., Schmitt, H.-D., and Arntz, E., Regulation of nitrate uptake in green algae and duckweeds — effect of starvation and induction, in *Biology of Inorganic Nitrogen and Sulfur,* Bothe, H., and Trebst, A., Eds., Springer-Verlag, Basel, 1981.

130. Thibaud, J. B. and Grignon, C., Mechanism of nitrate uptake in corn roots, *Plant Sci. Lett.,* 22, 279, 1981.

131. Kotyk, A., Critique of coupled vs. non-coupled transport of nonelectrolytes, *Proc. 5th Winter School Biophys. Membr. Transport,* Vol. 2, Agricultural University of Wrocław, 1979, 50.

132. Kotyk, A., Sources of energy for membrane transport in yeasts, in *Frontiers of Bioinorganic Chemistry and Molecular Biology,* Ananchenko, S. N., Ed., Pergamon Press, New York, 1980, 417.

133. Cambraia, J. and Hodges, T. K., ATPases of plasma membranes of oats roots, in *Plant Membrane Transport: Current Conceptual Issues,* Spanswick, R. M., Lucas, W. J., and Dainty, J., Eds., Elsevier, Amsterdam, 1980, 211.

134. Radin, J. W., Differential regulation of nitrate reductase induction in roots and shoots of cotton plants, *Plant Physiol.,* 55, 178, 1975.

135. Aslam, M. and Oaks, A., Comparative studies on the induction and inactivation of nitrate reductase in corn roots and leaves, *Plant Physiol.,* 57, 572, 1976.

136. Saglio, P. H. and Pradet, A., Soluble sugars, respiration, and energy charge during aging of excised maize root tips, *Plant Physiol.,* 66, 516, 1980.

137. Ullrich-Eberius, C. I., Novacky, A., Fischer, E., and Luttge, U., Relationship between energy-dependent phosphate uptake and the electrical membrane potential in Lemna gibba G1, *Plant Physiol.,* 67, 797, 1981.

138. Lin, W., Potassium and phosphate uptake in corn roots, Further evidence for an electrogenic H^+/K^+ exchanger and an OH^-/P antiporter, *Plant Physiol.,* 63, 952, 1979.

139. Tischner, R., The regulation of the nitrate metabolism in *Chlorella sorokiniana, Ber. Dtsch. Bot. Ges.,* 94, 635, 1981.

140. Tischner, R. and Lorenzen, H., Nitrate uptake and reduction in *Chlorella* — characterization of nitrate uptake in nitrate-grown and nitrogen-starved *Chlorella sorokiniana,* in *Biology of Inorganic Nitrogen and Sulfur,* Bothe, H. and Trebst, A., Eds., Springer-Verlag, Basel, 1981.

141. MacKown, C. T., Volk, R. J., and Jackson, W. A., Nitrate assimilation by decapitated corn root systems: effects of ammonium during induction, *Plant Sci. Lett.,* 24, 295, 1982.

142. Rufty, T. W., Jr., Jackson, W. A., and Raper, C. D., Jr., Inhibition of nitrate assimilation in roots in the presence of ammonium: the moderating influence of potassium, *J. Exp. Bot.,* 33, 1122, 1982.

143. Munn, D. A. and Jackson, W. A., Nitrate and ammonium uptake by rooted cuttings of sweet potato, *Agron. J.,* 70, 312, 1978.

144. Jackson, W. A., Nitrate acquisition and assimilation by higher plants: processes in the root sytem, in *Nitrogen in the Environment,* Nielsen, D. R. and MacDonald, J. G., Eds., Academic Press, New York, 1978, 45.

145. Minotti, P. L., Williams, D. C., and Jackson, W. A., Nitrate uptake and reduction as affected by calcium and potassium, *Soil Sci. Soc. Am. Proc.,* 32, 692, 1968.

146. Jackson, W. A., Flesher, D., and Hageman, R. H., Nitrate uptake by dark-grown corn seedlings: some characteristics of apparent induction, *Plant Physiol.,* 51, 120, 1973.

147. Blevins, D. G., Hiatt, A. J., and Lowe, R. H., The influence of nitrate and chloride uptake on expressed sap pH, organic acid synthesis, and potassium accumulation in higher plants, *Plant Physiol.,* 54, 82, 1974.

148. Ezeta, F. N. and Jackson, W. A., Nitrate translocation by detopped corn seedlings, *Plant Physiol.,* 56, 148, 1975.

149. Breteler, H., Hänisch Ten Cate, C. H. and Nissen, P., Time course of nitrate uptake and nitrate reductase activity in nitrogen-depleted dwarf bean, *Physiol. Plant.,* 47, 49, 1979.

150. Breteler, H. and Hänisch Ten Cate, C. H., Fate of nitrate during initial nitrate utilization by nitrogen-depleted dwarf bean, *Physiol. Plant.,* 48, 292, 1980.

151. Hänisch Ten Cate, C. H. and Breteler, H., Role of sugars in nitrate utilization by roots of dwarf bean, *Physiol. Plant.,* 52, 129, 1981.

152. Chang, N. K. and Jung, G. H., Nitrate reductase activity and nitrate uptake in rice seedling root, *Korean J. Bot. Shing'mul Hakhoe Ji,* 24, 159, 1981.

153. Breteler, H. and Luczak, W., Utilization of nitrite and nitrate by dwarf bean, *Planta,* 156, 226, 1982.

154. Doddema, H. and Otten, H., Uptake of nitrate by mutants of *Arabidopsis thaliana,* disturbed in uptake or reduction of nitrate, *Physiol. Plant.,* 45, 339, 1979.

155. Neyra, C. A. and Hageman, R. H., Nitrate uptake and induction of nitrate reductase in excised corn roots, *Plant Physiol.,* 56, 692, 1975.

156. Tompkins, G. A., Jackson, W. A., and Volk, R. J., Accelerated nitrate uptake in wheat seedlings: effects of ammonium and nitrite pretreatments and of 6-methylpurine and puromycin, *Physiol. Plant.,* 43, 166, 1978.

157. Hänisch Ten Cate, C. H. and Breteler, H., Effect of plant growth regulators on nitrate utilization by roots of nitrogen-depleted dwarf bean, *J. Exp. Bot.,* 33, 37, 1982.

158. Gotham, I. J. and Rhee, G-Y., Comparative kinetic studies of nitrate-limited growth and nitrate uptake in phytoplankton in continuous culture, *J. Phycol.,* 17, 309, 1981.

159. Braaksma, F. J. and Feenstra, W. J., Nitrate reduction in the wildtype and a nitrate reductase deficient mutant of *Arabidopsis thaliana, Physiol. Plant.,* 54, 351, 1982a.

160. Crom, W. J., Internal factor regulating nitrate and chloride influx in plant cells, *J. Exp. Bot.,* 24, 328, 1973.

161. Delmer, D. P., Dimethylsulfoxide as a potential tool for analysis of compartmentation in living plant cells, *Plant Physiol.,* 64, 623, 1979.

162. Hageman, R. H., Reed, A. J., Femmer, R. A., Sherrard, J. H., and Dalling, M. J., Some new aspects of the in vivo assay for nitrate reductase in wheat *(Triticum aestivum* L.) leaves, *Plant Physiol.,* 65, 27, 1980.

163. Aslam, M., Reevaluation of anaerobic nitrite production as an index for the measurement of metabolic pool of nitrate, *Plant Physiol.,* 68, 305, 1981.

164. Martinoia, E., Heck, U., and Wiemken, A., Vacuoles as storage compartments for nitrate in barley leaves, *Nature (London,)* 289, 292, 1980.

165. Granstedt, R. C. and Huffaker, R. C., Identification of the leaf vacuole as a major nitrate storage pool, *Plant Physiol.,* 70, 410, 1982.

166. Kinraide, T. B. and Etherton, B., H⁺-amino acid co-transport: influence of the acidic or basic character of the amino acids, *Plant Physiol.,* 63, S-12, 1979.

167. Kinraide, T. B. and Etherton, B., Electrical evidence for different mechanisms of uptake for basic, neutral, and acidic amino acids in oat coleoptiles, *Plant Physiol.,* 65, 1085, 1980.

168. Etherton, B. and Rubinstein, B., Evidence for amino acid-H⁺ co-transport in oat coleoptiles, *Plant Physiol.,* 61, 933, 1978.

169. Wann, M. and Raper, C. D., Jr., A dynamic model for plant growth: adaptation for vegetation growth of soybeans, *Crop Sci.,* 19, 461, 1979.

170. Brouwer, R., Nutritive influences on the distribution of dry matter in the plant, *Neth. J. Agric. Sci.,* 10, 399, 1962.

171. Thornley, J. H. M., A balanced quantitative model for root:shoot ratios in vegetative plants, *Ann. Bot.,* 36, 421, 1972.

172. Koster, A. L., Changes in metabolism of isolated root systems of soybean, *Nature (London),* 198, 709, 1963.

173. Koster, A. L., Enkele Aspecten von de Relatie Spriutwortel bij de Slekslofopname, Ph.D. dissertation, University of Leiden, The Netherlands, 1973.

174. Bowling, D. J. F., Graham, R. D., and Dunlop, J., The relationship between the cell electrical potential difference and salt uptake in the roots of *Helianthus annuus, J. Exp. Bot.,* 29, 135, 1978.

175. Chen, T. M. and Ries, S. K., Effect of light and temperature on nitrate uptake and nitrate reductase activity in rye and oat seedlings, *Can. J. Bot.,* 47, 341, 1969.

176. Swader, J. A., Stocking, C. R., and Lim, C. H., Light-stimulated absorption of nitrate by *Wolffia arrhiza, Physiol. Plant.,* 34, 335, 1975.

177. Hallmark, W. B. and Huffaker, R. C., The influence of ambient nitrate, temperature, and light on nitrate assimilation in sudan-grass seedlings, *Physiol. Plant.,* 44, 147, 1978.

178. Megel, K. and Viro, M., The significance of plant energy status for the uptake and incorporation of NH⁺₄ nitrogen by young rice plants, *Soil Sci. Plant Nutr.,* 24, 407, 1978.

179. Massimino, D., Andre, M., Richard, C., Daguenet, A., Massimino, J., and Vivoli, J., The effect of a day at low irradiance of a maize crop. I. Root respiration and uptake of N, P and K, *Physiol. Plant.,* 51, 150, 1981.

180. Pearson, C. J. and Steer, B. T., Daily changes in nitrite uptake and metabolism in *Capsicum annuum, Planta,* 137, 107, 1977.

181. Veen, B. W., The uptake of potassium, nitrate, water, and oxygen by a maize root system in relation to its size, *J. Exp. Bot.,* 28, 1389, 1977.

182. Hansen, G. K., Diurnal variation of root respiration rates and nitrate uptake as influence by nitrogen supply, *Physiol. Plant.,* 48, 421, 1980.

183. Wyche, R. D. and Rains, D. W., Nitrate absorption and acetylene reduction by soybeans during reproductive development, *Physiol. Plant.,* 47, 200, 1979.

184. Sasakawa, H. and Yamamoto, Y., Comparison of the uptake of nitrate and ammonium by rice seedlings, *Plant Physiol.,* 62, 665, 1978.

185. Smith, F. W. and Thompson, J. F., Regulation of nitrate reductase in excised barley roots, *Plant Physiol.,* 48, 219, 1971.

186. Lee, R. B., Inorganic nitrogen metabolism in barley roots under poorly aerated conditions, *J. Exp. Bot.,* 29, 693, 1978.

187. Aslam, M., Huffaker, R. C., Rains, D. W., and Rao, K. P., Influence of light and ambient carbon dioxide concentration on nitrate assimilation by intact barley seedlings, *Plant Physiol.,* 63, 1205, 1979.

188. Aslam, M. and Huffaker, R. C., *In vivo* nitrate reduction in roots and shoots of barley *(Hordeum vulgare* L.) seedlings in light and darkness, *Plant Physiol.,* 70, 1009, 1982.

189. Sahulka, J. and Lisa, L., The influence of sugars on nitrate reductase induction by exogenous nitrate or nitrite in excised *Pisum sativum* roots, *Biol. Plant.,* 20, 359, 1978.

190. Radin, J. W., Parker, L. L., and Sell, C. R., Partitioning of sugar between growth and nitrate reduction in cotton roots, *Plant Physiol.,* 62, 550, 1978.

191. Farrar, J. F., Respiration rate of barley roots: its relation to growth, substrate supply and the illumination of the shoot, *Ann. Bot. (London),* 48, 53, 1981.

192. Bowling, D. J. F., Evidence for an ion uptake controller in *Helianthus annuus*, in *Structure and Function of Plant Roots,* Brouwer, R., et al., Eds., Dr. W. Junk Publ, The Hague, 1981, 179.
193. Larsson, M., Ingemarsson, N., and Larsson, C-M., Photosynthetic energy supply for NO₃ assimilation in Scenedesmus, *Physiol. Plant.*, 55, 301, 1982.
194. Deanne-Drummond, C. E. and Clarkson, D. T., Effect of shoot removal and malate on the activity of nitrate reductase assayed in vivo in barley roots *(Hordeum vulgare* cv. Midas), *Plant Physiol.*, 64, 660, 1979.
195. Deanne-Drummond, C. E., Clarkson, D. T., and Johnson, C. B., The effect of differential root and shoot temperature on the nitrate reductase activity, assayed *in vivo* and *in vitro* in roots of *Horedum vulgare* (barley). Relationship with diurnal changes in endogenous malate and sugar, *Planta,* 148, 455, 1980.
196. Schrader, L. E., Domska, D., Jung, P. E., Jr., and Peterson, L. A., Uptake and assimilation of ammonium-N and nitrate-N and their influence on the growth of corn *(Zea mays* L.), *Agron. J.*, 64, 690, 1972.
197. Warncke, D. D. and Barber, S. A., Ammonium and nitrate uptake by corn (*Zea mays* L.) as influenced by nitrogen concentration and NH₄/NO₃ ratio, *Agron. J.*, 65, 950, 1973.
198. Edwards, J. H. and Barber, S. A., Nitrogen uptake characteristics of corn roots at low N concentration as influenced by plant age, *Agron. J.*, 68, 17, 1976.
199. Oaks, A., Stulen, I., and Boesel, I. L., Influence of amino acids and ammonium on nitrate reduction in corn seedlings, *Can. J. Bot.*, 57, 1824, 1979.
200. Radin, J. W., Amino acid interactions in the regulation of nitrate reductase induction in cotton root tips, *Plant Physiol.*, 60, 467, 1977.
201. Oaks, A., Aslam, M., and Boesel, I., Ammonium and amino acids as regulators of nitrate reductase in corn roots, *Plant Physiol.*, 59, 391, 1977.
202. Oaks, A., Nitrate reductase in roots and its regulation, in *Nitrogen Assimilation of Plants*, Hewitt, E. J. and Cutting, C. V., Eds., Academic Press, New York, 1979, 217.
203. Dunn-Coleman, N. S., Tomsett, A. B., and Garrett, R. H., Nitrogen metabolite repression of nitrate reductase in *Neurospora crassa:* effect of the gln-1a locus, *J. Bacteriol.*, 139, 697, 1979.
204. Premakumar, R., Sorger, G. J., and Gooden, D., Physiological characterization of a *Neurospora crassa* mutant with impaired regulation of nitrate reductase, *J. Bacteriol.*, 144, 542, 1980.
205. Flores, E., Guerrero, M. G., and Losada, M., Short-term ammonium inhibition of nitrate utilization by *Anacystis nidulans* and other cyanobacteria, *Arch. Microbiol.*, 128, 137, 1980.
206. Herrero, A., Flores, E., and Guerrero, M. G., Regulation of nitrate reductase levels in the cyanobacteria *Anacystis nidulans, Anabaena* sp. strain 7119, and *Nostoc* sp. strain 6719, *J. Bacteriol.*, 145, 175, 1981.
207. Rigano, V., Di, M., Vona, V., Fuggi, A., and Rigano, C., Effect of L-methionine-DL-sulphoximine, a specific inhibitor of glutamine synthetase, on ammonium and nitrate metabolism in the unicellular alga *Cyanidium caldarium, Physiol. Plant.*, 54, 47, 1982.
208. Slayman, C. L., Energetics and control of transport in *Neurospora,* in *Water Relations in Membrane Transport in Plants and Animals,* Jungreis, A. M., Hodges, T. K., Kleinzeller, A., and Schultz, S. G., Eds., Academic Press, New York, 1977, 69.
209. Keifer, D. W. and Lucas, W. J., Potassium channels in *Chara corallina:* control and interaction with the electrogenic H⁺ pump, *Plant Physiol.*, 69. 781, 1982.
210. Rai, A. K., Kashyop, A. K., and Gupta, S. L., ATP-dependent uptake of nitrate in *Nostoc Muscorum* and inhibition by ammonium ions, *Biochim. Biophys. Acta,* 674, 78, 1981.
211. Ohmori, M., Ohmori, K., and Strotmann, H., Inhibition of nitrate uptake by ammonia in a blue-green alga, *Anabaena cylindrica, Arch. Microbiol.*, 114, 225, 1977.
212. Pistorius, E. K., Funkhouser, E. A., and Voss, H., Effect of ammonium and ferricyanide on nitrate utilization by *Chlorella vulgaris, Planta,* 141, 279, 1978.
213. Tischner, R. and Lorenzen, H., Nitrate uptake and nitrate reduction in synchronous *Chlorella, Planta,* 146, 287, 1979.
214. Creswell, R. C. and Syrett, P. J., Ammonium inhibition of nitrate uptake by the diatom, *Phaeodactylum Tricornutum, Plant Sci. Lett.*, A, 321, 1979.
215. Cresswell, R. C. and Syrett, P. J., Uptake of nitrate by the diatom *Phaeodactylyum tricornutum, J. Exp. Bot.*, 32, 19, 1981.
216. Serra, J. L., Llama, M. J., and Cadenas, E., Nitrate utilization by the diatom *Skeletonema costatum.* I. Kinetics of nitrate uptake, *Plant Physiol.*, 62, 987, 1978.
217. Serra, J. L., Llama, M. J., and Cadenas, E., Nitrate utilization by the diatom *Skeletonema costatum.* II. Regulation of nitrate uptake, *Plant Physiol.*, 62, 991, 1978.
218. Schloemer, R. H. and Garrett, R. H., Nitrate transport system in *Neurospora crassa, J. Bacteriol.*, 118, 259, 1974.
219. Pate, J. S., Transport and partitioning of nitrogeneous solutes, *Annu. Rev. Plant Physiol.*, 31, 313, 1980.

220. Olday, F. C., Barker, A. V., and Maynard, D. N., A physiological basis for different patterns of nitrate accumulation in cucumber and pea, *J. Am. Soc. Hortic. Sci.,* 101, 219, 1976.
221. Radin, J. W., A physiological basis for the division of nitrate assimilation between roots and leaves, *Plant Sci. Lett.,* 13, 21, 1978.
222. Hunter, W. J., Fahring, C. J., Olsen, S. R., and Porter, L. K., Location of nitrate reduction in different soybean cultivars, *Crop. Sci.,* 22, 944, 1982.
223. Kirkman, M. A. and Miflin, B. J., The nitrate content and amino acid composition of the xylem fluid of spring wheat throughout the growing season, *J. Sci. Food Agric.,* 30, 653, 1979.
224. Rufty, T. W., Jr., Volk, R. J., McClure, P. R., Israel, D. W., and Raper, C. D., Jr., Relative content of nitrate and reduced N in xylem exudate as an indicator of root reduction of concurrently absorbed $^{15}NO_3$, *Plant Physiol.,* 69, 166, 1982.
225. Crafts-Brandner, S. J. and Harper, J. E., Nitrate reduction by roots of soybean (*Glycine max* L. Merr.) seedlings, *Plant Physiol.,* 69, 1298, 1982.
226. Radin, J. W., Contribution of the root system to nitrate assimilation in whole cotton plants, *Aust. J. Plant Physiol.,* 4, 811, 1977.
227. Butz, R. G. and Jackson, W. A., A mechanism for nitrate transport and reduction, *Phytochemistry,* 16, 409, 1977.
228. Heimer, Y. M. and Filner, P., Regulation of the nitrate assimilation pathway in cultured tobacco cells, *Biochim. Biophys. Acta,* 230, 363, 1971.
229. Goldsmith, J., Livoni, J. P., Norberg, C. L., and Segel, I. H., Regulation of nitrate uptake in Penicillium chrysogenium, *Plant Physiol.,* 52, 362, 1973.
230. Orebamjo, T. O. and Stewart, G. R., Some characteristics of nitrate reductase induction in *Lemna minor* L., *Planta,* 17, 1, 1974.
231. Rao, P. K. and Rains, D. W., Nitrate absorption by barley. II. Influence of nitrate reductase activity, *Plant Physiol.,* 57, 59, 1976.
232. Buczek, J., Kowalinska, E., and Kuczera, K., Reduction of nitrates in *Cucumis sativus* L. seedlings. I. Influence of tungsten and vanadium on absorption and reduction of nitrates, *Acta Soc. Bot. Pol.,* 49, 259, 1980.
233. Warner, R. L., Lin, C. J., and Kleinhofs, A., Nitrate reductase-deficient mutants in barley, *Nature (London),* 269, 406, 1977.
234. Warner, R. L. and Kleinhofs, A., Nitrate utilization by nitrate reductase-deficient barley mutants, *Plant Physiol.,* 67, 740, 1981.
235. Kleinhofs, A., Kuo, T., and Warner, R. L., Characterization of nitrate reductase deficient mutants, *Mol. Gen. Genet.,* 177, 421, 1980.
236. Oh, J. Y., Warner, R. L., and Kleinhofs, A., Effect of nitrate reductase deficiency upon growth, yield and protein in barley, *Crop Sci.,* 20, 487, 1980.
237. King, J. and Khanna, V., A nitrate reductase-less variant isolated from suspension cultures of *Datura innoxia* (Mill.), *Plant Physiol.,* 66, 632, 1980.
238. Feenstra, W. J. and Jacobsen, E., Isolation of a nitrate reductase deficient mutant of *Pisum sativum* by means of selection for chlorate resistance, *Theor. Appl. Genet.,* 58, 39, 1980.
239. Sosa, F. M., Ortega, T., and Barea, J. L., Mutants from *Chlamydomonas reinhardii* affected in their nitrate assimilation capability, *Plant Sci. Lett.,* 11, 51, 1978.
240. Nichols, G. L., Shehata, S. A. M., and Syrett, P. J., Nitrate reductase deficient mutants of *Chlamydomonas reinhardii.* Biochemical characteristics, *J. Gen. Microbiol.,* 108, 79, 1978.
241. Huffaker, R. C. and Rains, D. W., Factors influencing nitrate acquisition by plants: assimilation and fate of reduced nitrogen, in *Nitrogen in the Environment,* Nielsen, D. R. and MacDonald, J. G., Eds., Academic Press, New York, 1978, 1.
242. Ashley, D. A., Jackson, W. A., and Volk, R. J., Nitrate uptake and assimilation by wheat seedlings during initial exposure to nitrate, *Plant Physiol.,* 55, 1102, 1975.
243. Roldan, J. M., Verbelen, J.-P., Bulter, W. L., and Tokuyasu, K., Intracellular localization of nitrate reductase in *Neurospora crassa, Plant Physiol.,* 70, 872, 1982.
244. Smarrelli, J., Jr. and Campbell, W. H., NADH dehydrogenase activity of higher plant nitrate reductase, *Plant Sci. Lett.,* 16, 139, 1979.
245. Ekés, M., Ultrastructural demonstration of ferricyanide reductase (diaphorase) activity in the envelopes of plastids of etiolated barley *(Hordeum vulgare)* leaves, *Planta,* 151, 439, 1981.
246. Vaughn, K. C. and Duke, S. O., Histochemical localization of nitrate reductase, *Histochemistry,* 72, 191, 1982.
247. Läuchli, A., Spurr, A. R., and Epstein, E., Lateral transport of ions into the xylem of corn roots. II. Evaluation of a stelar pump, *Plant Physiol.,* 48, 118, 1971.
248. Läuchli, A., Pitman, M. G., Luttge, U., Kramer, D., and Ball, E., Are developing xylem vessels the site of ion exudation from root to shoot? *Plant Cell Environ.,* 1, 217, 1978.
249. Pitman, M. G., Ion transport into the xylem, *Annu. Rev. Plant Physiol.,* 28, 71, 1977.

250. Läuchli, A. and Epstein, E., Lateral transport of ions into the xylem of corn roots. I. Kinetics and energetics, *Plant Physiol.*, 48, 111, 1971.

251. Läuchli, A., Kramer, D., Pitman, M. G., and Lüttge, U., Ultrastructure of xylem parenchyma cells of barley roots in relation to ion transport to the xylem, *Planta*, 119, 85, 1974.

252. Läuchli, A., Kramer, D., and Stezler, R., Ultrastructure and ion localization in xylem parenchyma cells of roots, in *Membrane Transport in Plants*, Zimmerman, U. and Dainty, J., Eds., Springer-Verlag, Basel, 1974b, 363.

253. Leonard, R. W. and Hotchkiss, C. W., Plasma membrane-associated adenosine triphosphatase activity of isolated cortex and stele from corn roots, *Plant Physiol.*, 61, 175, 1978.

254. Schaefer, N., Wildes, R. A., and Pitman, M. G., Inhibition of p-fluorophenylalanine of protein synthesis and of ion transport across the roots in barley seedlings, *Austr. J. Plant Physiol.*, 2, 61, 1975.

255. Pitman, M. G., Lüttge, U., Laüchli, A., and Ball, E., Action of abscisic acid on ion transport as affected by root temperature and nutrient status, *J. Exp. Bot.*, 25, 147, 1974.

256. Pitman, M. G., Wildes, R. A., Schaefer, N., and Wellfare, D., Effect of azetidine 2 carboxylic acid on ion uptake and ion release to the xylem of excised barley roots, *Plant Physiol.*, 60, 240, 1977.

257. Pitman, M. G. and Wellfare, D., Inhibition of ion transport in excised barley roots by abscisic acid, relation to water permeability of the roots, *J. Exp. Bot.*, 29, 1125, 1978.

258. Pitman, M. G., Wellfare, D., and Carter, C., Reduction of hydraulic conductivity during inhibition of exudation from excised maize and barley roots, *Plant Physiol.*, 67, 802, 1981.

259. Bowling, D. J. F., Release of ions to the xylem in roots, *Physiol. Plant.*, 53, 392, 1981.

260. Morgan, M. A., Volk, R. J., and Jackson, W. A., Flurophenylalanine-induced alterations in nitrate uptake, reduction and translocation by maize roots, *Plant Physiol.*, 77, 718, 1985.

261. Waldron, J. C., Nitrogen compounds transported in the xylem of sugar cane, *Aust. J. Plant Physiol.*, 3, 415, 1976.

262. Frost, W. B., Belvins, D. G., and Barnett, N. M., Cation pretreatment effects on nitrate uptake, xylem exudate, and malate levels in wheat seedlings, *Plant Physiol.*, 61, 323, 1978.

263. Blevins, D. G., Hiatt, A. J., Lowe, R. H., and Leggett, J. E., Influence of K on the uptake, translocation, and reduction of nitrate by barley seedlings, *Agron. J.*, 70, 393, 1978.

264. Belvins, D. G., Barnett, N. M., and Frost, W. B., Role of potassium and malate in nitrate uptake and translocation by wheat seedlings, *Plant Physiol.*, 62, 784, 1987.

265. Touraine, B. and Grignon, C., Potassium effect on nitrate secretion into the xylem of corn roots, *Physiol. Veg.*, 20, 23, 1982.

266. Huber, S. C. and Moreland, D. E., Co-transport of potassium and sugars across the plasmalemma of mesophyll protoplasts, *Plant Physiol.*, 67, 163, 1981.

267. Touraine, B. and Grignon, C., Energetic coupling of nitrate secretion into the xylem of corn roots, *Physiol. Veg.*, 20, 33, 1982.

268. Cooil, B. J., Accumulation and radial transport of ions from potassium salts by cucumber roots, *Plant Physiol.*, 43, 158, 1974.

269. Cooil, B. J., Characteristics of radial solution flow in roots of cucumber *(Cucumis sativus* L.), *Ann. Bot.*, 38, 1043, 1947.

270. Shaner, D. L. and Boyer, J. S., Nitrate reductase activity in maize *(Zea mays* L.) leaves. I. Regulation by nitrate flux, *Plant Physiol.*, 58, 499, 1976.

271. Beevers, L. and Hageman, R. H., Nitrate and nitrite reduction, in *The Biochemistry of Plants, A Comprehensive Treatise,* Vol. 5, Amino Acids and Derivatives, Miflin, B. J., Ed., Academic Press, New York, 115, 1980.

272. Srivastava, H. S., Regulation of nitrate reductase activity in higher plants, *Phytochemistry,* 19, 725, 1980.

273. Naik, M. S., Abrol, Y. P., Nair, T. V. R., and Ramarao, C. S., Nitrate assimilation: its regulation and relationship to reduced nitrogen in higher plants, *Phytochemistry,* 21, 495, 1982.

274. Campbell, W. H. and Smarrelli, J., Jr., Nitrate reductase: properties and regulation, in *Biochemical Basis of Plant Breeding,* Vol. 2, Neyra, C., Ed., CRC Press, Boca Raton, Fla., 1986, chap. 1.

275. Yoneyama, T., ^{15}N studies on the *in vivo* assay of nitrate reductase in leaves: occurrence of underestimation of the activity due to dark assimilation of nitrate and nitrite, *Plant Cell Physiol.*, 22, 1507, 1981.

276. Blevins, D. G., Lowe, R. H., and Staples, L., Nitrate reductase in barley roots under sterile, low oxygen conditions, *Plant Physiol.*, 57, 458, 1976.

277. Shaner, D. L. and Boyer, J. S., Nitrate reductase activity in maize, *(Zea mays* L.) leaves. II. Regulation by nitrate flux at low leaf water potential, *Plant Physiol.*, 58, 505, 1976.

278. Morrison, S. L., Huffaker, R. C., and Guidara, C. R., Light interaction with nitrate reduction, in *Genetic Engineering of Symbiotic Nitrogen Fixation and Conservation of Fixed Nitrogen,* Lyons, J. M., Valentine, R. C., Phillips, D. A., Rains, D. W., and Huffaker, R. C., Eds., Plenum Press, New York, 1981, 547.

279. Reed, A. J. and Hageman, R. H., Relationship between nitrate uptake, flux, and reduction and the accumulation of reduced nitrogen in maize *(Zea Mays* L.). I. Genotypic variation, *Plant Physiol.,* 66, 1179, 1980.

280. Reed, A. J. and Hageman, R. H., Relationship between nitrate uptake, flux and reduction and the accumulation of reduced nitrogen in maize *(Zea mays* L.). II. Effect of nutrient nitrate concentration, *Plant Physiol.,* 66, 1184, 1980.

281. Steer, B. T., Rhythmic nitrate reductase activity in leaves of *Copsicum annum* L. and the influence of kinetin, *Plant Physiol.,* 57, 928, 1976.

282. Steer, B. T., Integration of photosynthetic carbon metabolism and nitrogen metabolism on a daily basis, in *Proc. Conf. Head at the "Limburgs Universitair Centrum",* Marcelle, R., Clijsters, H., and Van Poucke, M., Eds., Dr. W. Junk Publ., The Hague, 1979, 309.

283. Moore, R. and Black, C. C., Jr., Nitrogen assimilation pathways in leaf mesophyll and bundle sheath cells of C4 photosynthesis plants formulated from comparative studies with *Digitaria sanguinalis* (L.) Scop, *Plant Physiol.,* 64, 309, 1979.

284. Neyra, C. A. and Hageman, R. H., Pathway of nitrate assimilation in corn *(Zea mays* L.) leaves. Cellular distribution of enzymes and energy sources for nitrate reduction, *Plant Physiol.,* 62, 618, 1978.

285. Aslam, M., Huffaker, R. C., and Delwiche, C. C., Reduction of nitrate and nitrite in barley leaves in darkness, in *Genetic Engineering of Symbiotic Nitrogen Fixation and Conservation of Fixed Nitrogen,* Lyons, J. M., Valentine, R. C., Phillips, D. A., Rains, D. W., and Huffaker, R. C., Eds., Plenum Press, New York, 1981, 533.

286. Meeker, G. B., Purvis, A. C., Neyra, C. A., and Hageman, R. H., Uptake and accumulation of nitrate as a major factor in the regulation of nitrate reductase activity in corn *(Zea mays* L.) leaves: effects of high ambient CO_2 and malate, in *Mechanisms of Regulation of Plant Growth,* Bieleski, R. L., Ferguson, A. R. and Cresswell, M. M., Eds., Bulletin 12, The Royal Society of New Zealand, Wellington, New Zealand, 1974, 49.

287. Reed, A. J. and Canvin, D. T., Light and dark controls of nitrate reduction wheat *(Triticum aestivum* L.) protoplasts, *Plant Physiol.,* 69, 508, 1982.

288. Kondo, T., Correlation between potassium uptake rhythm and nitrate uptake rhythm in *Lemna gibba* G3, *Plant Cell Physiol.,* 23, 909, 1982.

289. Moll, R. H., Kamprath, E. J., and Jackson, W. A., The potential for genetic improvement in nitrogen use in maize, *Proc. Annu. Corn and Sorghum Industry-Research Conf.,* 37, 163, 1983.

Chapter 5

USE OF PHYSIOLOGICAL TRAITS, ESPECIALLY THOSE OF NITROGEN METABOLISM FOR SELECTION IN MAIZE

J. H. Sherrard, R. J. Lambert, F. E. Below, R. T. Dunand, M. J. Messmer, M. R. Willman, C. S. Winkels, and R. H. Hageman

TABLE OF CONTENTS

I. INTRODUCTION

Conventional plant breeding techniques have resulted in increased plant yield largely through selection based on observable economic performance, modification of canopy, tolerance to disease, adaptability to improved managerial practices, especially higher fertility, and suitability for new agricultural areas. While yields of crop plants throughout the world have risen steadily during the past four decades, it is now being questioned as to whether this rate of increase can be maintained using these same procedures.[1] However, it is not a matter of exhausting the genetic material available for use in selection programs, as plant breeders are generally confident that sufficient "genetic reserves" exist for this purpose. The continued progress in selection from an open-pollinated cultivar for high and low grain protein and oil after 76 generations and the success of reverse selection (reverse high and reverse low for both components) for 40 generations attests to the existence of sufficient genetic variability to allow for long-term progress in maize.[2] Other studies indicate that approximately 60% of the steady increase in grain yields of maize over the past 50 years has been due to improvement in genetic composition.[3,4] The progress from selection for both chemical composition and grain yield indicates that existing germplasm contains sufficient genetic variability to permit continued progress for the foreseeable future. Thus, the problem is not the lack of genetic variability but how to identify and exploit a variable physiological or biochemical trait that is causally related to productivity.

In the U.S., it has been estimated that replacement of open pollinated varieties with hybrid maize has increased yields by 15 to 30%.[5] Since the introduction of hybrid maize in Illinois there has been a steady and continued increase in the average grain yield for the state,[6] with a record of 134 bu/A in 1982. An altered canopy, improved photosynthesis, a responsiveness to N fertilizer, disease resistance, and stalk quality have been associated with these increased yields. Changes in the physiological traits, if any, have been achieved indirectly via selection for yield. Those observations raise the questions of what the upper limit of productivity on a national scale is and whether useful physiological or biochemical traits can be identified and exploited.

Murata and Matsushima[7] estimated the upper theoretical yield of rice to be twice the record yield, and Wittwer[8] indicated a threefold difference for maize. The record yield of maize (325 bu/A, nonirrigated, Herman Warsaw, Saybrook, Ill.) is 2.4-fold greater than the record 1-year average yield for the state of Illinois. These observations indicate that the production potential of the current maize genotypes is adequate for the current status of managerial skills, or that current cultivars are not well adapted for the environments under which much of the crop is grown.

According to Hageman et al.[9] or Mahon,[10] a biochemical or physiological trait should meet the following requirements:

1.   Genetic variability for the trait must be present in the parental material.
2.   The trait must be highly heritable.
3.   An accurate and preferably simple method for measuring the trait is needed.
4.   The trait should show a positive association with yield.

Genetic variability and heritability for major and important physiological or biochemical enzymes or systems such as nitrogen metabolism, photosynthesis, nitrogen fixation and dark respiration have been reported for numerous plant species.[11-15] In these cases, adequate, but not ideal, methods for measurement of the trait were developed. Unfortunately, under field conditions the association between the various physiological or biochemical traits and yields have not been consistent or dramatic enough to lead to widespread commercial utilization.

Tollenaar[16] stated that the importance of current photoassimilation in influencing grain yield is affected by the relationship between availability of assimilate and sink size. Discussion and expansion of this statement reveals the complexities and problems involved in attempting to identify traits related to nitrogen metabolism or photosynthesis that can be used in the selection of a superior cultivar. Availability and metabolism of nitrogen have been accorded a major role in the establishment and maintenance of the photosynthetic canopy and in the establishment of sink size.[7,17] For rice, Murata and Matsushima[7] stated that application of excess nitrogen at certain stages of plant development resulted in excessive leaf-area expansion which was inversely correlated with photosynthetic capacity. In cereals, kernel growth is dependent not only on the initial kernel dry weight and response to substrate supply, but also on the availability of current and stored photosynthate during the grain-filling period.[18] Evans[19] pointed out that in many cases, it is difficult to determine whether source or sink is limiting grain yields, because of the involvement and interactions of other factors (capacity to translocate assimilants, deficiencies of water or nutrients at different stages of growth, and environmental conditions). Tanaka and Yamaguchi[20] concluded that at the *present yield levels* grain yield of maize is limited by sink capacity and of rice by source capacity. Based on the findings of Allison and Watson[21] and other researchers that maize stalks retain appreciable amounts of carbohydrates during the grain filling period, Tollenaar[16] concluded that sink capacity was most frequently the factor limiting grain yield of maize grown in the central Corn Belt of the U.S. For maize grown under the short season environments encountered in Canada, Hume and Campbell[22] concluded that the source was the limiting factor. Environmental and cultural conditions affect the amount of carbohydrates retained by the maize stalk at the time of grain black layer formation. Stalks contained 65 g carbohydrates per stalk when the maize hybrid B73 × Mo17 was grown by Below et al.[23] in 1980 (39,500 plants ha^{-1}, 91 q ha^{-1} grain yield) and 3 g carbohydrate per stalk when grown by Swank et al.[24] on similar plots in 1981 (59,250 plants ha^{-1}, 96 q ha^{-1} grain yield). Computed from the loss of carbohydrates from the vegetation during the grain-filling period of five maize hybrids, Swank et al.[24] estimated that current photosynthesis supplied 97% (average for the five hybrids) of the ear dry weight. Stalks of all hybrids retained appreciable levels (70% of that present at 5 days prior to anthesis) of carbohydrates by time of grain maturity. In contrast, the average amount of nitrogen in the ear derived from current assimilation was 31% for these same five hybrids. Nitrogen loss from the leaves of all hybrids was nearly linear throughout the grain-filling period and this loss of leaf nitrogen is considered to be associated with, if not the cause of, the decline in photosynthetic activity during grain fill.

Cessation of kernel growth, frequently induced by stress conditions, does not appear to be directly related to plant carbohydrate level. During such stress periods stalk carbohydrate level often increases[25] and a high concentration of carbohydrate is usually present in both stalk and kernels.[26,27] Genetic differences exist among genotypes of cereal crops in the ability to sustain kernel growth through mobilization of plant reserves.[28,29] However, the reduction in grain production under stress conditions during initiation and early kernel development is more likely a result of an inadequate supply of available nitrogen. Kernel initiates, pollen, and developing embryo all possess proportionately high levels of nitrogen. The importance of nitrogen during the early stages of grain development can also be inferred from work of Ingle et al.[30] where changes were measured in the composition of developing maize kernels. This work indicated that endosperm cell division was essentially completed within 28 days following pollination. The association between endosperm cell number and grain yield of wheat has more recently been established.[31,32]

Frey[33] reported that altered assimilate supply (as affected by defoliation or thinning at 50% silking) influenced maize yield by altering both kernel number per plant and final kernel weight. Reduction in the level of irradiation during flowering was more detrimental to the grain yield of maize than comparable reduction of light during the grain-filling period.[34,35] While these data could indicate that photosynthate supply affects sink capacity, it is equally possible that the treatments adversely affected the supply of reduced nitrogen, as light energy also enhances nitrate assimilation. In maize, decreased irradiation depressed nitrate assimilation to a greater extent than carbohydrate metabolism[36] Nitrate reductase is more sensitive to decreased leaf water potential than either photosynthesis or respiration, and nitrate reductase is more sensitive to elevated temperatures than either phosphoenolpyruvate or ribulose 1,5-biphosphate carboxylase.[6,37]

The relationship between nitrate reductase activity and grain yield is further confounded by other physiological and biochemical processes occurring during ear development. In maize, ear removal adversely affected photosynthesis and nitrate assimilation to the same extent.[38a] Ear removal had little effect on nitrogen remobilization, indicating that remobilization is controlled independently of ear sink. Based on the correlation found between the accumulation of zein and grain yield, Tsai et al.[38b] have suggested that protein accumulation might be an important factor in grain development in maize. In a study of five maize genotypes, Swank et al.[24] found that the rate of nitrogen accumulation in the developing ear was not related to the rate of dry matter accumulation. This study, however, does not preclude the possibility that dry matter accumulation is dependent upon concurrent transport of nitrogen and carbohydrate to the developing grain. Excessive remobilization of nitrogen from the vegetation could result in high grain nitrogen and/or a decrease in assimilate supply through degradation of the leaf photosynthetic apparatus. Reed et al.[39] concluded from their study of four maize genotypes that leaf proteolytic activity was more closely related to accumulation of grain nitrogen than leaf nitrate reductase activity.

The identification of a useful physiological or biochemical trait is complicated because the direct effect of the trait is buffered by the complex metabolic systems that control the sequence of plant processes culminating in yield. Consequently, the effects of a given trait on productivity can be expected to vary in the same genotype grown under different environments or among genotypes. It is also possible that a useful selection trait may have a short-term effectiveness before some other trait becomes the limiting factor. For example, Austin et al.[40] have suggested that plant breeders are close to exhausting the usefulness of dry matter distribution as a trait to increase yields, although in the past, this trait was successfully used to enhance productivity.

Yield has been used so successfully as a selection trait because it effectively integrates the interactions of the metabolic processes and the environmental effects on these processes. Yield measurements are accurate and simple. Yield is also the primary standard for evaluation of a physiological or biochemical trait. However, yield as a selection trait provides only an empirical evaluation of integration that occurred and provides no information on the metabolic processes that underlie productivity.

One advantage of selection based on physiological or biochemical traits that characterize major metabolic systems, even when unsuccessful with respect to yield, is that such studies provide knowledge of the action and interaction of the processes. Although separate attempts to use photosynthetic or nitrate reductase activities for the selection of high yielding maize cultivars were unsuccessful, these studies provide indirect support for the need to simultaneously evaluate the effect of more than one metabolic trait. This view is consistent with the proposed need for a balanced metabolic system.[11] More knowledge of the requirements of the plant at critical stages of development and the metabolism that meets these requirements is needed if metabolic systems that enhance yield are to be successfully incorporated into the plant by classical or DNA-transfer procedures.

II. PHOTOSYNTHETIC ACTIVITY AS A SELECTION TRAIT

The high concentration of carbon in all plant material (40 to 45% C, dry weight basis) dictates a major role for photosynthetic activity in crop productivity. Photosynthetic activity as a selection trait meets the first three of the four proposed guidelines. Genetic variability in photosynthetic rates per unit leaf area has been reported for several crop plants.[19,41,42] Heritability of light-saturated photosynthetic rate has been found high for *Lolium perenne*[43] and maize.[42] Numerous methods have been developed for measurement of the trait. However, a positive relationship between photosynthetic rates and yield of various crops is seldom observed.

With wheat, Evans and Dunstone[44] have found modern high-yielding cultivars have lower photosynthetic rates than the ancestral types. Similarly, Austin et al.[45] found photosynthesis per unit area of the flag leaf was highest for diploids, intermediate for triploids, and lowest for the hexaploid *Triticum aestivum*. Pearlman et al.[46] found no relationship between grain yields and photosynthesis for several spring wheat cultivars. In contrast to these reports, Kulshrestha and Tsunoda[47] found that cultivars with the "Norin 10" dwarfing genes had significantly higher photosynthetic rates per unit leaf area than tall wheats lacking these genes. The "Norin 10" cultivars, under favorable conditions of growth, have a greater potential for grain production than the tall varieties. Earlier, Vishitani et al.[48] had noted that barley with a semi-dwarf gene had higher photosynthetic rates per unit leaf area than the cultivar of normal height.

With maize, Moss and Musgrave[42] successfully selected for divergent photosynthetic rates for five generations; however, they found no correlative response in grain yield. Fakorede and Mock[49] reported that recurrent selection for grain yield had no effect on carbon exchange rates for a population cross [BSSS(R) X BSCB1 (R)]. They concluded that increases in grain yield resulted from increases in translocation efficiency and in duration of transport of assimilate to the grain. Crosbie et al.[50] found that selection for carbon exchange rates resulted in an actual advance of about 5% per cycle with realized heritability of 0.26 to 0.33. In their study, one cycle of selection for higher carbon exchange rates had no effect on plant height, tillering, or grain yield.

With soybeans, Curtis et al.[51] and Dornhoff and Shibles[41] noted a variation of carbon exchange rates among cultivars; however, there was no consistent relationship between photosynthesis and seed yield. Harrison et al.[52] found that apparent canopy photosynthesis was significantly correlated with seed yield in a number of soybean lines derived from two crosses. Wells et al.[53] found correlations (r ranged from 0.59 to 0.66) between the canopy apparent photosynthesis integrated over the time period, full-pod stage to maturity, and seed yield for 16 soybean cultivars. They concluded that differences in canopy photosynthesis partially accounted for differences in seed yield.

With rice, Murata[17] and Osada[54] have shown differences in photosynthetic rates per unit leaf area among cultivars. With six cultivars grown at two levels of N fertility (normal and high), Murata[17] found no relationship between the mean photosynthetic rates during the grain-filling period and yield. In response to added N fertilizer, the average increase of photosynthetic rates for five of the cultivars was 5% and the average increase in grain yield was 24%. The sixth cultivar had a 7% increase in photosynthesis and a 7% decrease in grain yield.

Numerous explanations have been given for the inconsistent relations between photosynthetic rates and plant productivity; however, none have been completely satisfactory. A negative relationship between photosynthetic rates and leaf size has frequently been observed among some crop plants,[55] but this relation was not detected in barley.[56]

Rates of photosynthesis per unit leaf area are affected by variations in leaf thickness, mesophyll cell thickness and size, and leaf chlorophyll content.[43] Rates of photosynthesis per mesophyll cell have been shown to be less variable than rates per unit leaf area.[57] However, among wheat species no relation was observed between cell size and leaf or grain size, and a low negative correlation was found between cell size and photosynthetic rates.[58] Bowes et al.[59] found that the light intensity under which the leaf was grown altered photosynthetic rates and level of RuBP carboxylase activity. With rice cultivars that differed in chlorophyll content, Murata[17] found a correlation between photosynthetic rate and chlorophyll content at low but not at saturating light levels. This finding indicates that differences in diurnal photosynthetic activity patterns could exist among field-grown cultivars that differed in chlorophyll content. It can also be suggested that the yields currently obtained are so far below the theoretical potential yield that the capacity for photosynthetic production is not often the limiting factor.

"Inefficient" dark respiration, where energy is expended on processes not directly contributing to growth (protein turnover, integrity of organelles and membranes, and ion uptake) could contribute to the lack of relationship between photosynthesis and productivity. This "maintenance" requirement for respiration may vary both between species and genotypes and throughout plant development. Apel and Tschapa[60] have demonstrated variation among wheat cultivars in the efficiency of respiration of the ear during grain filling, with the grain contributing 30 to 50% of the respiration of the entire ear. Morgan and Austin[61] suggested that the maintenance component of respiration during grain filling of wheat was substantial and that selection based upon efficient use of energy might be useful. With ryegrass, the rate of dark respiration of mature leaves and roots can vary considerably, even between lines of the same cultivar.[62] In differentiating between the maintenance component of respiration and the "biosynthetic" component (that provides energy predominantly for cell division and expansion) Robson[63] has found the maintenance component to be higher (two-fold) in undefoliated ryegrass swords with high leaf-area index. Presumably, productivity would be increased by reducing the energy requirements for plant maintenance. This has been recognized in a number of studies where a negative relationship exists between dark respiration of leaves or roots, and productivity.[62,64] Respiration measurements in these studies could be influenced by both maintenance and biosynthetic components, although Penning de Vries[65] has suggested that the genetic variation in the biosynthetic component of respiration is likely to be small. This was later confirmed by Wilson,[66] who found that selection for low respiration families of *Lolium perenne* increased biomass by 7% while having little effect on growth during early stages of development when the biosynthetic components are likely to be proportionately larger. Wilson[67] reported increases of 20 to 30% in production of digestible organic matter by selection for reduced dark respiration rate. These lines of ryegrass with low respiration used less carbon for maintenance than lines with normal respiration.[68,69]

In addition to establishing the potential usefulness of respiration rate as a selection trait, the experiment of Wilson[69] and Robson[68] provide another reason for the lack of a consistent relationship between photosynthesis and grain yield. The desirability of measuring both respiration and photosynthesis is indicated. An estimate of the respiratory component could be obtained by the difference between canopy photosynthesis and dry matter accumulation (net photosynthesis) when both components were expressed on comparable terms. Equipment for measurement of canopy photosynthesis is available, even for field-grown maize; however, the equipment is expensive, and the procedure complex. Although selection based on dry weight accumulation might effectively integrate both photosynthesis and respiration, there is a need for separate evaluation of these two processes.

III. NITROGEN METABOLISM AS A SELECTION TRAIT

The economic return to the producer from the increased use of fertilizer N is a compelling reason for considering N metabolism as a critical component in crop productivity. In Illinois over the past three decades, increased grain yield of maize tended to parallel increased application of fertilizer N.[6] The grain yield per unit of N applied decreased rapidly between 1950 and 1965 and very slowly from 1965 to 1979.[13] The data up to 1965 indicate the inadequacy of indigenous soil N for maize production, while the data for 1965 to 1979 shows a proportional small increase in grain yield for each unit of N applied. Duvick[4] found that the more recently released maize hybrids were more productive than hybrids released at an earlier date when both were grown on moderately high levels of fertilizer N. Welch[70] found that grain yield of newly released hybrids decreased dramatically when the annual application of fertilizer N was omitted. For rice, high yields are possible only under high levels of N supply, and plant characteristics that confer high yielding abilities are often associated with responsiveness to fertilizer N.[7]

A. Nitrate Reductase

Because nitrate is the predominant form of soil N available to crop plants under nonpaddy condition, and nitrate reductase activity (NRA) is substrate (NO_3) inducible and the rate-limiting step in the assimilation of nitrate,[6,37] it seemed logical to test NRA of the leaves as a simple selection criterion. Genetic variability in NRA was shown to exist for maize[71-73] and for wheat.[74,75] The work of Warner et al.[76] and Schrader et al.[73] established that the trait was heritable and subject to manipulation by classical maize breeding techniques. As with photosynthetic rates NRA has not been consistently related to yields of grain or grain protein for maize[77,78] or wheat.[74,75] In the initial evaluation of NRA in maize,[77] the two high NRA hybrids were subjectively known to outyield the two low NRA hybrids. However, in comparable plot tests, for two consecutive years there was no difference in yield among the four hybrids, although substantial and consistent differences in NRA were reconfirmed.

For a given wheat variety, NRA of the plant integrated over time showed a high degree of correlation with the actual amount of N accumulated by the plant.[75,79] Comparisons made among varieties showed no close correlation between NRA and yield of grain or grain protein. Dalling et al.[80] found highly significant correlations between canopy NRA and actual accumulation of N for five wheat varieties. The correlation between NRA and yield of grain or grain protein was nonsignificant; however, when corrections were made for differences in harvest index for N, NRA and grain protein production were significant (r = 0.95).

Because of the characteristics of nitrate reductase, it was initially assumed that NRA (measured by either in vivo or in vitro procedures) provided an estimate of the current availability of reduced N to the plant. Current data show that this assumption is not valid among diverse genotypes, although for a given genotype, NRA provides a reasonable estimate of the availability of reduced N. *In situ* accumulation of reduced N is influenced not only by the level of the enzymes but by the availability of nitrate at the induction and assimilation sites[81,82] and availability of reductant and metabolites for regeneration of reductant.[79,83,84] Reed and Hageman[85,86] found that among maize genotypes a given flux of nitrate did not result in the induction of the same amount of NRA.

In 1974, a divergent, phenotypic recurrent selection program with NRA as the sole selection trait was initiated to more directly assess the effect of variation in level of NRA on N metabolism and productivity of material having a common background.[87] Super stiff-stalk synthetic maize was used as the parental material because several su-

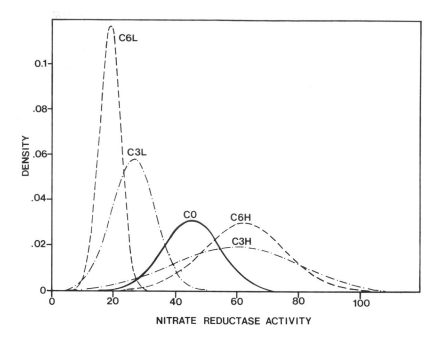

FIGURE 1.   The density (frequency/class interval) distribution obtained
for the original selection from the maize super stiff-stalk synthetic parental material (Cycle 0) and from the divergently selected high and low nitrate reductase strains of Cycles 3 and 6 (C3H, C6H, C3L, and C6L).

perior inbreds had been developed from this synthetic. For each selection cycle for
both high and low strains, the first and second leaf above the ear of 200 selfed plants
were removed at 10 and 20 days after anthesis, respectively, and assayed for NRA. For
each plant, the NRA of the two sampling dates was integrated over time and modified
to reflect the rate of change in activity between the two samplings. A 20% selection
intensity was used to select the 40 high and 40 low selfed plants. An equal number of
kernels from each of the 40 ears of both strains were bulked and planted. Within each
selection strain, the plants were intermated to produce the respective high and low
strains. By intermating the plants in a Hawaiian winter nursery, one cycle of selection
could be completed each year. Initially, it was assumed that increasing NRA in the
leaves after anthesis would be associated with extended leaf area duration was well as
with increased amounts of reduced N in the plant.

   The density (frequency/class interval) distribution and related data obtained for the
original selection from the parental material and from divergent (high and low NRA)
strains of Cycles 3 and 6 are presented in Figure 1 and Table 1. Cycles 3 and 6 of the
low NRA strains had different distributions than the comparable high strains. The
change in means of the low strains was much greater over the six cycles of selection
(45.4 vs. 19.0) than for the high strains (45.4 vs. 62.6). The standard deviations for the
high strains were three to four times greater than for the low strains (Table 1). The
estimates of skewness (asymmetry) and of kurtosis (peakedness or flatness) were different for Cycles 3 and 6 in both high and low strains (Table 1). The changes in distribution indicate that both cycles may be approaching the limits of selection. The large
standard deviation of the high NRA strains may indicate a greater environmental effect
on level of enzyme activity. Alternatively, the selection for NRA may have indirectly
selected for differential nitrate uptake as the plants were grown on soils with marginal
levels of available nitrogen (100 kg N ha^{-1}, annually). Consequently genetic variability
for NRA per se could have been decreased and the indirect selection effect enhanced.

Table 1

FREQUENCY DISTRIBUTION DATA FOR THE HIGH AND LOW NITRATE
REDUCTASE STRAINS DEVELOPED BY PHENOTYPIC RECURRENT
SELECTION FOR CYCLES 0, 3, AND 6 IN THE MAIZE SYNTHETIC STIFF
STALK

	NRA		SD[a]		Skewness		Kurtosis	
	High	Low	High	Low	High	Low	High	Low
Cycle 0	45.4		12.9		+0.65[b]		+0.33	
Cycle 3	60.6	26.6	20.4	6.9	+1.35[b]	+1.15[b]	+1.94[b]	+2.71[b]
Cycle 6	62.6	19.0	13.2	3.4	+0.33	+0.11	−0.24	−0.46

Note: NRA = nitrate reductase activity expressed as μmol NO_2^- accumulated hr^{-1} leaf $^{-1}$.

[a] SD = standard deviation.
[b] Significantly different from a standardized normal distribution.

This explanation is supported to some extent by the differences in standard deviation between Cycles 3 and 6 of the divergent selected strains (3.5 for low and 7.2 for high; Table 1). These data strongly support the conclusion that divergent selection for NRA was very effective.

Initial comparisons of level of NRA and grain yield of the initial cycles of selection were summarized by Dunand,[87] Sherrard et al.,[88] and Messmer,[89] and a more extensive evaluation of Cycles 2, 4, and 6 of the high and low strains and the parental material was made in 1982. The 1982 data confirms Messmer's results[89] that the divergence in NRA among the strains has been significant and continuous (Figure 2A). After six cycles of selection, activity of the high NRA strain increased 52% over cycle 0 (parental material), while activity in the low strain decreased by 38%, relative to Cycle 0. For both strains, the level of NRA declined throughout the season (Figure 2B). These data show that selection for NRA affected the level of enzyme activity and the rate of decline of activity during the grain-filling period and confirm the heritability of the enzyme and that the level of activity could be altered by classical procedures.

With respect to correlated responses, the low NRA strains had higher concentration of leaf chlorophyll than the high NRA strains, when assays were made at 21 days after anthesis (Figure 2C). The seasonal profiles (Figure 2D) show that initially the concentration of chlorophyll was similar for both the high and low NRA strains, but that the high NRA selections lost chlorophyll from the leaf at a faster rate than the low NRA selections during the grain-filling period. For the Cycle-4 material (Figure 2D), the concentration of leaf chlorophyll declined 80% for the high and 65% for the low NRA strain during grain fill. Similar seasonal patterns were exhibited by Cycle-2 and -6 materials. Whether the maintenance of higher levels of leaf chlorophyll by the low NRA selections during the grain-filling period was associated with a similar maintenance of photosynthetic activity was not determined.

Specific leaf weights (gram dry weight cm^{-2}) were greater in the high than in the low NRA strains (Figure 2E). The trends over the three cycles (2, 4, and 6) show a positive response in specific leaf weight with selection for high NRA and a slight decrease in specific leaf weight with selection for low NRA. Differences in leaf weight and leaf area were observed but they exhibited no consistent pattern in response to selection. The seasonal profiles of specific leaf weights of a representative cycle show that divergence between the two strains was maintained throughout the sampling period. While the metabolic significance of this correlated response is not clear, the data show that selection based on NRA can affect an anatomical trait.

Between 5 days before and 31 days after anthesis, the low NRA strains consistently

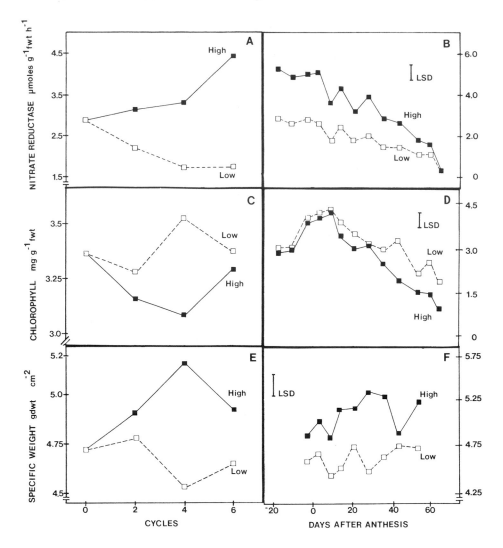

FIGURE 2.   Effect of divergent selection for nitrate reductase activity of maize leaves, and the correlative responses of two physiological traits and of the selected strains and parental material. Assays were made with the first leaf above the ear for the parental material (Cycle 0) and Cycles 2, 4, and 6 of the divergently selected (high and low) nitrate reductase activity populations. Data of A, C, and E show the divergence between the cycles when measurements were made at 21 days after anthesis. Data of B, D, and F show the seasonal divergence for Cycle 4. Data presented are representative for season and cycles. Specific leaf weights are to be multiplied by 10^{-3} (e.g., 0.005 g dry wt cm^{-2}).

had higher leaf nitrate content than the high NRA strains (Table 2). Selection for high NRA did not markedly decrease the leaf nitrate content throughout the season, relative to the Cycle 0 material. In contrast, selection for low NRA was associated with higher leaf nitrate content during the first half of the grain filling period, relative to Cycle 0. Relative to the second cycle, successive selections for low NRA have not resulted in additional increases in leaf nitrate content, which indicates that factors other than level of nitrate reductase are affecting amounts of nitrate in the leaf. The higher content could indicate that the low NRA strains absorb more nitrate than the high NRA strains. Alternatively, and more consistent with the data of nitrogen accumulated by the total plant, the inability of the low NRA strain to induce high levels of nitrate reductase, in relation to the levels of substrate (nitrate) present, may result in a tem-

Table 2

EFFECT OF DIVERGENT SELECTION FOR LEAF
NITRATE REDUCTASE ACTIVITY (NRA) ON
NITRATE CONTENT OF MAIZE

Cycle[a]	Days after anthesis						
	−5	8	18	25	31	37	45
	mg NO_3^--N plant part^{-1}						
0	64.4	34.3	33.7	20.9	15.1	30.4	18.0
2 High	67.3	27.0	31.9	20.7	16.6	31.2	25.9
2 Low	88.6	44.3	39.8	37.3	29.1	29.4	21.3
4 High	62.9	41.8	29.8	17.6	30.0	20.1	33.2
4 Low	86.7	36.7	39.9	41.4	47.1	30.9	26.3
6 High	65.2	43.1	23.9	22.8	24.9	22.8	24.6
6 Low	90.5	44.1	36.1	35.0	34.2	24.7	31.8

Note: Assays were made on composite samples of all leaves at intervals during the 1982 growing season.

[a] The material evaluated was parental material (Cycle 0) and Cycles 2,
4, and 6 of the divergently selected (high and low with respect to leaf
NRA) populations.

porary excess of nitrate. Thus, more nitrate would likely be available for storage in the
vacuoles, where it would be available for assimilation (albeit slowly). It is conceivable
that selection for leaf NRA may have inadvertently selected for differences in nitrate
partitioning.

In response to selection, there was no effect on reduced (N) content of stover or
grain, with the exception that Cycle 6 low NRA strain had a significantly lower amount
of grain nitrogen (Table 3). With the successive cycles of selection, the high and low
NRA strains exhibited divergent trends in concentration of stover reduced N while the
concentrations of nitrogen in the grain remained relatively constant. There was no
relation between the relative amounts (seasonal averages) of NRA and the reduced N
accumulated by the plants (total above-ground parts) of the different cycles, as judged
by the efficiency indexes (Table 3). These results show that selection for increased leaf
NRA is not an effective way of increasing reduced N content of the plant and indicate
that at reasonable-to-high levels, the enzyme is not the factor that limits nitrate assimilation. However, the low seasonal mean NRA of Cycle (-6) low strain (1.7 μmol NO_2
accumulated g^{-1} fresh leaf wt hr^{-1}) was associated with a sharp decrease in yield of
grain nitrogen and grain. This raises the possibility that below some threshold level,
the enzyme can become a limiting factor. The adverse affect of limiting N on grain but
not stover production is consistent with the observation of Krantz and Chandler[90] that
increasing soil nitrogen from deficient to sufficient levels increased grain weight three-
to fourfold more than stover weight.

The evaluation of the parental material and six cycles of divergently selected strains
(composite of seven experiments over 5 years) shows a significant depression of grain
yield only for cycle-6 low NRA strain (Figure 3). Similar to reduced N content, there
was no relationship between grain yield and the relative amounts (seasonal averages)
of NRA among the seven entries.

Selection of additional cycles is underway to determine whether further decreases in
the level of NRA can be achieved and associated with progressive decreases in grain
yield and grain protein and adverse effects on accumulation of dry weight and reduced-
N by the vegetation.

Table 3

DRY WEIGHT AND REDUCED NITROGEN CONTENT OF GRAIN AND STOVER AND NITRATE REDUCTASE EFFICIENCY INDEXES OF THE PARENTAL MAIZE MATERIAL (CYCLE 0) AND CYCLES 2, 4, AND 6 POPULATION FROM A DIVERGENT (HIGH AND LOW) RECURRENT PHENOTYPIC SELECTION FOR POSTANTHESIS LEAF NITRATE REDUCTASE ACTIVITY

Selection cycle	Dry weight (g plant⁻¹)				Reduced N (g plant⁻¹)				NRA efficiency index Total plant N/seasonal mean of NRA ($g\ N/\mu mol\ NO_3^-\ g^{-1}fwt\ hr^{-1}$)	
	Grain		Stover		Grain		Stover			
	High	Low	High	Low	High	Low	High	Low	High	Low
0	174		132		2.9		1.1		1.4	
2	180	183	136	137	3.0	3.1	1.0	1.2	1.8	2.1
4	176	176	148	158	3.0	3.2	1.2	1.3	1.2	2.4
6	175	120	146	158	2.9	2.1	1.3	1.2	1.0	1.9
LSD (0.05)	31		NS		0.6		NS		—	

Note: Three plants per plot were harvested at time of grain maturity. NRA were seasonal mean values (weekly measurements made between 20 days prior to 60 days after anthesis.

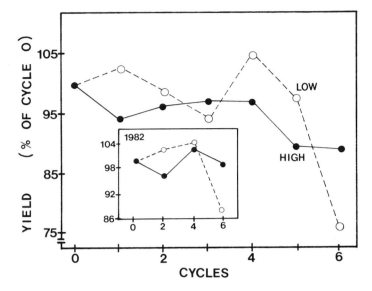

FIGURE 3.   Effect of divergent selection for postanthesis level of nitrate reductase activity of maize leaves on grain yield. The data were compiled from seven separate experiments: 1976 (one), 1978 (one), 1979 (two), 1980 (one), and 1982 (two). The yield of each selected cycle was expressed relative to the parental material (Cycle 0, 100%) for each experiment and then averaged over experiments. Not all cycles were evaluated in every experiment; however, the minimum evaluation involved three experiments (2 years). The number of replicate plots for each trial ranged from five to ten. Cycle 0 yields ranged from 58.3 (1979) to 90.6 (1982) q ha^{-1}. Data for 1982 (insert) was for parental material (Cycle 0) and Cycles 2, 4, and 6 of the divergently selected (high and low) nitrate reductase population. In contrast to the data of Table 2, the yields were obtained by harvesting the middle row of yield plots.

B. Nitrogen Content or Concentration

Mass-selection breeding methods have been shown to be effective for improving grain yields, pest resistance, and other traits in maize populations. Hallauer and Miranda[91] have summarized the progress for different selection methods used for population improvement in maize. They reported a range in gain per cycle from 13 to 18% for the half-sib family selection method. This method should be useful for the identification of physiological or biochemical traits, because an estimation of "narrow-sense" heritability and genetic correlations can be obtained for each trait for each cycle of selection. It is conceivable that the selection response of the half-sib family method may be improved if an appropriate trait(s) could be identified.

In 1980, a selection program was initiated using half-sib families of an Illinois version of stiff-stalk synthetic maize (RSSSC). One hundred half-sib families were evaluated under field conditions (replicate in block design) for grain yields and various nitrogen traits. The families were grown in single row plots, ten families per block with two replicates. Each row was 5.23 m long and 0.76 m apart. Final stand count was 56,586 plants ha^{-1}. At anthesis, three plants per plot were sampled. The plants were divided into leaves and stalks and the parts assayed for content and concentration of nitrate N, reduced N, and total N. At grain maturity, the remaining plants were harvested for grain yield.

From the initial planting in 1980, 20 of the highest yielding families (20% selection intensity) were selected to consitiute the high-yield strain. Similarly, 20 families with the highest composite score for high grain yields and high reduced-N content of the leaves were selected to constitute the high yield - high reduced N strain. Remnant seed

Table 4

MEANS, COEFFICIENTS OF VARIATION, AND
HERITABILITY ESTIMATES OF VARIOUS TRAITS, AND
GRAIN YIELD FOR THE HALF-SIB MAIZE FAMILIES OF
ILLINOIS STIFF-STALK SYNTHETIC GROWN AT URBANA,
ILLINOIS IN 1980 AND 1981

Plant part and trait	Mean (%)		G.C.V. (%)[a]		Heritability (%)[b]	
	1980	1981	1980	1981	1980	1981
Leaf						
Dry weight (g plant^{-1})	54.0	45.6	6.1	7.6	60[c]	59[c]
Reduced N (g plant^{-1})	1.5	1.3	5.9	8.1	44[c]	47[c]
Reduced N (mg g^{-1} dry wt)	27.2	28.4	4.8	2.8	35[c]	27
Stalk						
Dry weight (g plant^{-1})	98.1	65.1	4.2	9.5	24	56[c]
Reduced N (g plant^{-1})	0.9	0.7	6.2	7.9	49[c]	37[c]
Reduced N (mg g^{-1} dry wt)	8.8	10.9	5.3	4.1	43[c]	20
Grain yield (g plant^{-1})	116	127	8.4	14.8	31[c]	69[c]

Note: Evaluations in 1980 were on 100 families from the synthetic and in 1981 on 100
families from the high grain yield high leaf reduced-N content strain.

[a]   Genetic coefficient of variation.
[b]   Narrow-sense heritability estimates on an individual plot basis.
[c]   Significantly different from zero at the 5% probability level.

of the selected families for each strain were grown, separately, in the Hawaiian winter
nursery, and plants intermated to generate the seeds that constitute cycle one material
for each strain. The selection process was then repeated for both strains.

Portions of the data obtained from these selection experiments are presented in Table 4. In 1980, the heritability estimates for leaf dry weight and leaf reduced N content
were markedly higher than for grain yield. Heritability estimates for leaf reduced N
concentration and grain yield were similar, and the relative amounts of genetic variability of this trait were double that of grain yield. In 1980, the genetic correlation
between leaf reduced N content and grain yield was 0.70 ± 0.19. The correlation between leaf dry weight and grain yield was 0.52 ± 0.20, while none of the other four
traits were significantly related to grain yield. Although the 1980 data indicated a potential usefulness of the criteria of yield and leaf reduced-N content, this was not supported by the 1981 data (Table 4). Heritability estimates and the genetic coefficients of
variation were greater for grain yield than for the six traits, especially the nitrogen
traits. None of the six traits was significantly correlated with grain yield. These data
indicate that environmental effects markedly influenced the estimates of heritability
and genetic coefficient of variation of grain yield as well as the biological selection
traits.

The effect of the environment is evident by the variation in mean dry weights and
reduced-N contents and concentrations of leaves and stalks (Table 4). Although the
plants were smaller and had less reduced-N per plant in 1981 than in 1980, the reduced-
N concentration in both leaves and stalks was higher. These differences in plant growth
and composition are attributed to differences in rainfall and temperature between the
two years. The plants experienced severe drought stress in 1980 during the early phases
of grain fill. These results show the difficulties encountered in utilizing physiological
or biochemical traits in mass-selection procedures. Patience is mandatory. However,

in 1982 a preliminary field evaluation of the parental material (Cycle 0) and Cycle 1 of the two selected strains (grain yield and grain yield - leaf reduced-N content) were conducted. Cycle 1 of the yield strain outyielded Cycle 0 by 10% (752 kg ha^{-1}). Cycle 1 of the yield - reduced N strain yielded 7% (501 kg ha^{-1}) more than Cycle 0. Additional cycles of selected material need to be obtained and concurrently evaluated under different environments to validate the effects of the selection trait.

Muruli and Paulsen[92] successfully used mass selection for grain yield and half-sib families in maize to identify a biochemical trait associated with grain yield. N-inefficient and -efficient strains were selected from the highest yielding families grown at high (200 kg N ha^{-1}) and low (0) nitrogen, respectively. These selected strains were then evaluated for yield at 0, 50, 100, and 200 kg N ha^{-1}. At low levels of N, the N-efficient selection outyielded both the parental material and the N-inefficient strains, but were not responsive at the high levels of N. The N-inefficient strains yielded most at the high levels of N. They concluded that N-use efficiency could be improved through selection. Of numerous N traits measured, the amount of reduced N accumulated by the plant (stover plus grain) between silking and maturity was closely associated with the progress in yield relative to parental material for both classes of selected strains.

In addition to the selection programs previously described, physiological and biochemical traits have potential use in cultivar development. The potential use of certain leaf nitrogen traits of S_2 lines and hybrids of maize for development of inbreds was investigated by Messmer.[93]

From the selection phase of the ongoing phenotypic divergent recurrent selection program with NRA (Section III. A), 200 plants, classed as Cycle 3 high NRA, were selfed and subsequently each plant was evaluated as previously described. Ears selected from the 75 plants having highest leaf NRA provided the seed for the 75-Cycle 3 selfed-once ($C3S_1$) high NRA families. In 1978, these 75 families were evaluated for grain yield. Remnant seed of the same 75 families were also grown in 1978 and plants within each family were self-pollinated to provide a group of 75-$C3S_2$ high NRA families. Based upon rankings for leaf NRA obtained in 1977 and yield in 1978 the $C3S_2$ families were ranked, separately, for NRA and yield. From these rankings, two classes of $C3S_2$ families were identified: high NRA-high yield (families with highest yields and NRA [combined score]) and high NRA-low yield (families with lowest yield and highest NRA [combined score]). In a similar manner, the two classes of $C3S_2$ families, low NRA-high yields and low NRA-low yields, were derived from the divergently selected 200 cycle 3 low NRA plants. A total of five families (ears) were selected for each of these four classes of material.

The classes of families, $C4S_2$ high NRA-high yield, high NRA-low yield, low NRA-high yield, and low NRA-low yield, were derived in a similar manner from the divergently selected Cycle 4 plants.

In 1979, the 20 families (five families from each of the four classes) were crossed with four commercial inbreds (H96, Mo17, B73, and N28) to produce 80 different categories of test cross-hybrid seed. Reciprocal crosses were made in all cases and no distinction was made with respect to male or female parent. The choice of the commercial inbred seed was arbitrary; however, H96 and Mo17 had previously been shown to have relatively high and low levels of leaf NRA, respectively.[87] In 1980, all 80 test cross-hybrids (five families of each of the four $C3S_2$ classes × four commercial inbreds) were grown in the field (two replicate plots per hybrid). The plants in each replicate plot were evaluated for leaf NRA and reduced-N concentration. At 10 and 20 days after anthesis, three leaves taken from above the ear and three leaves from just below the ear, respectively, from six representative plants from each replicate plot were composited for assay. Grain yield was also determined on the center (yield) row. The ranking of these 80 hybrids for leaf reduced-N concentration gave the basis for an addi-

Table 5

NITRATE REDUCTASE ACTIVITY AND REDUCED NITROGEN
CONCENTRATION OF THE LEAF LAMINAE AND GRAIN YIELD OF 32
TEST CROSS-MAIZE HYBRIDS PREVIOUSLY CLASSIFIED FOR HIGH
AND LOW CONCENTRATION OF ENZYME AND NITROGEN

Genotypic class[a]	1980 NRA ($\geqslant$mol $NO_3^- \cdot g^{-1}$fwt$\cdot hr^{-1}$)	Reduced nitrogen (mg g^{-1} dry wt)			Grain yield (kg ha^{-1})			
		1980	1981	Mean	1980	1981	1982	Mean
HRN	3.92	34.9	35.9	35.4	7796	7957	10030	8549
LRN	3.82	32.7[b]	35.1[b]	33.9[b]	6686	7322[b]	9022[b]	7676[b]
HNRA	4.35	33.7	35.1	34.3	7260	7434	9173	7956
LNRA	3.46[b]	33.9	35.9[b]	34.9[b]	7422	7846[c]	9879[c]	8382[b]
HNRA-HRN	4.34	34.8	35.6	35.2	7586	7436	9482	8168
HNRA-LRN	4.36	32.5	34.5	33.5	6934	7432	8864	7743
LNRA-HRN	3.65	35.1	36.2	35.6	8006	8479	10597	9027
LNRA-LRN	3.28	32.8	35.6	34.2	6838	7212	9180	7743
LSD 0.05[d](1)	0.29	0.7	0.5	0.3	—	452	683	288
(2)	0.41	0.6	0.5	0.2	—	516	774	268

Note: Hybrids were grown at a high level of fertility in 1980, 1981, and 1982 at Urbana, Ill.

[a] Genotypic classes are high and low leaf lamina total reduced nitrogen concentration (HRN and LRN, respectively), high and low leaf lamina nitrate reductase activity (HNRA and LNRA, respectively), and combinations of RN and NRA classes (HNRA-HRN, etc.).
[b] Significant at the 0.01 probability level.
[c] Significant at the 0.05 probability level.
[d] LSD 0.05 denotes Fisher's Least Significant Difference for comparison of (1) HRN vs. LRN means within NRA classes, and (2) RN class means in different NRA classes.

tional classification as follows. From each group of five test cross-hybrids (e.g., one group of five test crosses would be the five high NRA high yield C3S$_2$ families × B73, that were grown in replicate plots in stratified block design) one test cross hybrid with the highest and one with the lowest leaf reduced-N concentration were selected from each stratified block. The evaluation for leaf reduced-N concentration established eight classes of material for each of the four inbred testers. These classes for the inbred B73 are as follows:

Inbred × C3S$_2$ families	Reduced-N class
B73 × high NRA high yield	High
B73 × high NRA high yield	Low
B73 × high NRA low yield	High
B73 × high NRA low yield	Low
B73 × low NRA high yield	High
B73 × low NRA high yield	Low
B73 × low NRA low yield	High
B73 × low NRA low yield	Low

Using appropriate remnant seed, these 32 test cross-hybrids were grown under field conditions (five replicate plots, with other conditions as described in Section III.B) and evaluated for concentration of leaf reduced-N and yield in 1981 and yield in 1982.

The data, averaged over the inbred testers for the major genotypic classification for traits, are presented in Table 5. In 1980, the year of selection, there was significant difference in leaf reduced-N concentration for the HRN and LRN genotypic classes. These differences were associated with a 13% significant increase in grain yield. In 1981, the HRN class again had significantly higher leaf reduced-N and grain yield than the LRN class. When averaged over 3 years, the HRN hybrids had 10.5% higher grain yields than the LRN hybrids.

In 1980, the HNRA class of hybrids had 26% more nitrate reductase activity than the LNRA class of hybrids. These data indicate the classification of the S_2 lines for level of NRA effectively predicted the level of enzyme activity of the hybrids derived from the S_2 lines. However, this increased activity did not result in increased concentrations of reduced N in the leaves in either 1980 or 1981. In 1981 and averaged over both years the LNRA class had significantly more reduced N in the leaves than the HNRA class. Grain yields were significantly higher for the LNRA class in 1981, 1982, or averaged over the 3 years than for the HNRA class.

When the hybrids were grouped by dual traits, the LNRA-HRN class had the highest concentration of laminae reduced N and grain yields in all instances (Table 5). A comparison of the LNRA-LRN class with the HNRA-LRN class showed little effect of level of NRA on grain yield, while the comparison of HNRA-HRN vs. LNRA-HRN indicates a possible adverse effect of high enzyme activity on productivity.

Additional evidence for the apparent adverse effect of high levels of leaf NRA on grain yield was obtained in a separate experiment. The hybrids were produced by crossing the four major classes (based on leaf laminae NRA and grain yield [Section III.B]) of twice selfed (S_2) families (two families per class) from both Cycle 3 and Cycle 4 with the two inbred testers H96 and Mo17. Dunand[87] had shown that Mo17 had low levels of NRA relative to H96. These families (two families per class) were never ranked for concentration of reduced N in the leaf laminae. These data, averaged and grouped for statistical analysis on the basis of enzyme level without regard for yield classification, are presented in Table 6. Grain yields were significantly higher when H96 was crossed with either Cycle 3 or 4 S_2 families previously classified as low NRA lines than when crossed with Cycle 3 or 4 high NRA families. When the inbred tester was Mo17, grain yields were similar regardless of NRA classification. Hybrids of high NRA lineage had significantly higher NRA than hybrids of low NRA lineage. The effect of the NRA level of the inbreds was also evident, as the overall mean level of leaf NRA of all test cross-hybrids involved with H96 was 4.2 μmol NO_2 accumulated per gram fresh weight per hour vs. 2.9 for the Mo17 material. The reason for the reduction in grain yields of hybrids derived for the high NRA families is not clear and work is underway to extend and confirm these initial observations.

Yield components were measured for 8 of the 32 cross-hybrids grown in 1981 and 1982 as described previously (Section III. B, Table 5). The eight hybrids selected were the progeny of the inbred testers (H96 and Mo17) and the four $C3S_2$ families (Section III. B, Table 5) classed for high yield. These data, averaged over the two years and grouped by major classes, are presented in Table 7. The HRN class had higher grain yields, more kernels per plant, faster rates of kernel dry weight accumulation during the last half of grain fill and smaller kernels at maturity than the LRN class. Other comparisons showed that the HRN class matured earlier (lower grain moisture at harvest, 22.8 vs. 23.3%), and silked 2 days later than the LRN class. Most components were unaffected when the grouping was by NRA; however, HNRA had fewer kernels per plant and accumulated more grain dry weight during the first half of grain fill. When the hybrids were grouped by dual traits, the LNRA-HRN class had the highest grain yield (although not significant at the 0.05% level), the most kernels per plant, and the smallest kernels at maturity, relative to the other three classes. For all four classes, kernel number was negatively related to kernel weight (Table 7).

These data (Tables 5, 6, and 7) indicate that high concentrations of reduced N and

Table 6
GRAIN YIELD AND NITRATE REDUCTASE
ACTIVITY OF THE LEAF LAMINAE OF TEST
CROSS-MAIZE HYBRIDS OBTAINED BY CROSSING
TWO INBRED TESTERS AND CYCLE 3 AND CYCLE
4 S_2 FAMILIES DERIVED FROM A DIVERGENT
RECURRENT SELECTION PROGRAM FOR HIGH
AND LOW LEVELS OF NITRATE REDUCTASE
ACTIVITY

Genotypic class				
Inbred	Family cycle	Class NRA	Grain yield (kg/ha^{-1})	Leaf NRA (μmol NO$_2^-$ g^{-1} fwt hr^{-1})
H96	C3S$_2$	High	11,273	5.1[a]
H96	C3S$_2$	Low	12,091[a]	3.9
Mo17	C3S$_2$	High	10,852	3.5[a]
Mo17	C3S$_2$	Low	10,699	2.7
H96	C4S$_2$	High	11,686	4.8[a]
H96	C4S$_2$	Low	12,198[a]	3.0
Mo17	C4S$_2$	High	11,027	3.0[b]
Mo17	C4S$_2$	Low	11,021	2.4

Note: Although the S_2 lines were previously classified with respect to
high and low levels of enzyme and grain yield, the data were
grouped for analysis only on the basis of enzyme activity.

[a]   Denotes statistically significant differences at the 1% confidence level.
[b]   Denotes statistically significant differences at the 5% confidence level.

low levels of nitrate reductase activity in the leaf laminae during the first third of the grain-filling period (LNRA-HRN class) may be a useful criteria for the development of high-yielding maize hybrids. It should be pointed out that this may be applicable only to the genotypes utilized in these experiments. Currently it is not clear why LNRA as a trait should be beneficial, as the level of NRA should be a measure of the flux of nitrate into the leaf.[81] How can high levels of leaf reduced N be attained without high rates of flux of nitrate? Obviously, other factors are involved. The view that nitrogen has a major role in establishment and maintenance of the photosynthetic canopy[7] provides a possible explanation for the beneficial effects of high concentrations of reduced N in the leaf. The work of Christensen et al.,[38a] Swank et al.,[24] and unpublished data show that major amounts of reduced N, chlorophyll, and photosynthetic activity are concurrently lost from maize leaves throughout the grain filling period.

Assuming that the maintenance of high levels of leaf reduced N throughout the grain-filling period is the cause of increased productivity, these observations raise the following questions. How can the concentration of reduced N be increased and maintained in the leaves of maize during the grain-filling period? Could the rate and duration of grain fill exert a regulating effect on photosynthetic activity? What are the metabolic processes involved and the number of genes involved in the regulation of these processes? If the answers to these questions were known, it would be much easier to utilize either classical breeding procedures in conjunction with biochemical traits or genetic engineering to improve the productivity of maize hybrids.

Table 7

TWO-YEAR MEAN FOR GRAIN YIELD, KERNELS PER PLANT, KERNEL
WEIGHT AT TWO DATES, AND THE GAIN IN KERNEL WEIGHT BETWEEN
THESE DATES FOR EIGHT MAIZE HYBRIDS

Genotypic class[a]	Grain yield (kg/ha^{-1})	No. of kernels per plant	Maturity (g 100 kernels^{-1})	Kernel weight 35 days after anthesis (g 100 kernels^{-1})	Kernel weight gain 35 days after anthesis to maturity (g 100 kernels^{-1})
HRN	9972	672	26.5	18.1	8.4
LRN	8738[b]	632[b]	27.6[b]	20.1[b]	7.4[c]
HNRA	9313	644	27.3	19.5	7.8
LNRA	9397	660[c]	26.8	18.7[b]	8.0
HNRA-HRN	9875	653	27.8	18.0	9.7
HNRA-LRN	8752	635	26.9	21.0	5.9
LNRA-HRN	10070	691	25.2	18.1	7.1
LNRA-LRN	8723	630	28.3	19.1	9.0
LSD .05[d]	ns	18	0.9	0.5	1.0

Note: Hybrids were obtained by crossing Cycle 3 S_2 families classed as high and low for nitrate reductase activity, total reduced nitrogen concentration, and high yields with the inbred testers H96 and Mo17. Hybrids were grown at a high level of fertility in 1981 and 1982 at Urbana, Ill.

[a] Genotypic classes are high and low leaf lamina reduced nitrogen concentration (HRN and LRN, respectively), high and low leaf lamina nitrate reductase activity (HNRA and LRNA, respectively), and combinations of RN and NRA classes (HNRA-HRN, etc.).
[b] Significant at the 0.01 probability level.
[c] Significant at the 0.05 probability level.
[d] LSD 0.05 denotes Fisher's Least Significant Difference for comparing NRA-RN class means at 0.05 level.

REFERENCES

1. Wittwer, S. H., The next generation of agricultural research, *Science,* 199, 375, 1978.
2. Dudley, J. W., 76 generations of selection for oil and protein percentage in maize, *in Proc. Int. Conf. Quantitative Genetics,* Pollack, E., Kempthorne, O., and Bailey, T. B., Eds., Iowa State University Press, Ames, 1977, 459.
3. Cardwell, V. B., Fifty years of Minnesota corn production, *Agron. J.,* 74, 984, 1982.
4. Duvick, D. V., Genetic rates of gain in hybrid maize yields during the past forty years, *Maydica,* 22, 187, 1977.
5. Sprague, G. F., Alexander, D. E., and Dudley, J. W., Research Priorities in Plant Breeding, Special Report of the Department of Agronomy, University of Illinois, Urbana, 1978.
6. Hageman, R. H., Integration of nitrogen assimilation in relation to yield, in *Nitrogen Assimilation of Plants,* Hewitt, E. J. and Cutting, C. V., Eds., Academic Press, New York, 1979, 591.
7. Murata, Y. and Matsushima, S., Rice, in *Crop Physiology,* Evans, L. T., Ed., Cambridge University Press, London, 1975, 73.
8. Wittwer, S. H., The shape of things to come, in *The Biology of Crop Productivity,* Carlson, P. S., Ed., Academic Press, New York, 1980, 413.
9. Hageman, R. H., Lambert, R. J., Loussaert, D., Dalling, M., and Klepper, L. A., Nitrate and nitrite reductases as factors limiting protein synthesis, in *Genetic Improvement in Seed Protein,* Workshop Proceedings, National Academy of Sciences, Washington, D. C., 1976, 103.
10. Mahon, J. D., Limitations to the use of physiological variability in plant breeding, *Can. J. Plant Sci.,* 63, 11, 1983.
11. Hageman, R. H., Leng, E. R., and Dudley, J. W., A biochemical approach to corn breeding, *Adv. Agron.,* 19, 45, 1967.
12. Hobbs, S. L. A. and Mahon, J. D., Variation, heritability and relationship to yield of physiological characters in peas, *Crop Sci.,* 22, 773, 1982.

13. Sherrard, J. H., Lambert, R. J., Messmer, M. J., Below, F. E., and Hageman, R. H., Plant breeding for efficient use of nitrogen, in *Nitrogen in Crop Production,* Hauck, R. D., Ed., American Society of Agronomy, Madison, 1984, 363.

14. Vose, P. B. and Breese, E. L., Genetic variation in the utilization of nitrogen by ryegrass species *Lolium perenne* and *L. multiflorum, Ann. Bot.,* 28, 251, 1964.

15. Wilson, D., Breeding for morphological and physiological traits, in *Plant Breeding II,* Frey, K. J., Ed., Iowa State University Press, Ames, 1981, 233.

16. Tollenaar, M., Sink-source relationships during reproductive development in maize. A review, *Maydica,* 22, 49, 1977.

17. Murata, Y., Photosynthesis, respiration and nitrogen response, in *Mineral Nutrition of the Rice Plant,* Johns Hopkins Press, Baltimore, 1965, 388.

18. Thornley, J. H. M., Wheat grain growth: anthesis to maturity, *Aust. J. Plant Physiol.,* 6, 187, 1979.

19. Evans, L. T., The physiological basis of crop yield, in *Crop Physiology: Some Case Histories,* Evans, L. T., Ed., Cambridge University Press, London, 1975, 327.

20. Tanaka, A. and Yamaguchi, J., Dry matter production, yield components and grain yield of the maize plant, *J. Fac. Agric. Hokkaida Univ.,* 57, 72, 1972.

21. Allison, J. C. S. and Watson, D. J., The production and distribution of dry matter in maize after flowering, *Ann. Bot.,* 30(3), 5, 1966.

22. Hume, D. J. and Campbell, D. K., Accumulation and translocation of soluble solids in corn stalks, *Can. J. Plant. Sci.,* 52, 363, 1972.

23. Below, F. E., Christensen, L. E., Reed, A. J., and Hageman, R. H., Availability of reduced N and carbohydrates for ear development of maize, *Plant Physiol.,* 68, 1186, 1981.

24. Swank, J. C., Below, F. E., Lambert, R. J., and Hageman, R. H., Interaction of carbon and nitrogen metabolism in the productivity of maize, *Plant Physiol.,* 70, 1185, 1982.

25. Williams, W. A., Loomis, R. S., Duncan, W. G., Doverat, A., and Nunez, F., Canopy architecture at various population densities and the growth and grain yield of corn, *Crop. Sci.,* 8, 303, 1968.

26. Tollenaar, M. and Daynard, T. B., Effect of defoliation on kernel development in maize, *Can. J. Plant Sci.,* 58, 207, 1978.

27. Tollenaar, M. and Daynard, T. B., Dry weight, soluble sugar content and starch content of maize kernels during the early postsilking period, *Can. J. Plant. Sci.,* 58, 199, 1978.

28. Austin, R. B., Edrich, J. A., Ford, M. A., and Blackwell, R. D., The fate of dry matter, carbohydrates, and ^{14}C loss from leaves and stems of wheat during grain filling, *Ann. Bot.,* 41, 1309, 1977.

29. Blum, A., Paiarkova, H., Golan, G., and Mayer, N., Chemical dessication of wheat plants as a simulator of post-anthesis stress. I. Effects of translocation and kernel growth, *Field Crops Res.,* 6, 51, 1983.

30. Ingle, J., Beitz, D., and Hageman, R. H., Changes in composition during development and maturation of maize seeds, *Plant Physiol.,* 40, 835, 1965.

31. Brocklehurst, P. A., Factors controlling grain weight in wheat, *Nature (London),* 266, 348, 1977.

32. Gleadow, R. M., Dalling, M. J., and Halloran, G. M., Variation in endosperm characteristics and nitrogen content in six wheat lines, *Aust. J. Plant Physiol.,* 9, 539, 1982.

33. Frey, N. M., Dry matter accumulation in kernels of maize, *Crop Sci.,* 21, 118, 1981.

34. Earley, E. B., McIlrath, W. O., Seif, R. D., and Hageman, R. H., Effect of shade applied at different stages in corn *(Zea mays,* L.) production, *Crop Sci.,* 7, 151, 1967.

35. Prine, G. M., Critical period for ear development among different eartypes of maize, *Soil Crop Sci. Soc. Fla. Proc.,* 33, 27, 1973.

36. Knipmeyer, J. W., Hageman, R. H., Earley, E. B., and Seif, R. D., Effect of light intensity on certain metabolites of the corn plant *(Zea mays* L.), *Crop Sci.,* 2, 1, 1962.

37. Beevers, L. and Hageman, R. H., Nitrate and nitrite reduction, in *Aminoacids and Derivatives: The Biochemistry of Plants,* Miflin, B. J., Ed., 1980, 5, 115.

38a. Christensen, L. E., Below, F. E., and Hageman, R. H., The effects of ear removal on senescence and metabolism of maize, *Plant Physiol.,* 68, 1180, 1981.

38b. Tsai, C. Y., Huber, D. M., and Warren, H. L., Relationship of the kernel sink for N to maize productivity, *Crop Sci.,* 18, 399, 1978.

39. Reed, A. J., Below, F. E., and Hageman, R. H., Grain protein accumulation and the relationship between leaf nitrate reductase and protease activities during grain development in maize *(Zea mays* L.). I. Variation between genotypes, *Plant Physiol.,* 66, 164, 1980.

40. Austin, R. B., Bingham, J., Blackwell, R. D., Evans L. T., Ford, M. A., Morgan, C. L., and Taylor, M., Genetic improvements in winter wheat yields since 1900 and associated physiological changes, *J. Agric. Sci. (Cambridge),* 94, 675, 1980.

41. Dornhoff, G. M. and Shibles, R. M., Varietal differences in net photosynthesis of soybean leaves, *Crop. Sci.,* 10, 42, 1970.

42. Moss, D. N. and Musgrove, R. B., Photosynthesis and crop production, *Adv. Agron.,* 23, 317, 1971.

43. Wilson, D. and Cooper, P. J. Diallel analysis of photosynthetic rate and related leaf characters among contrasting genotypes of *Lolium perenne, Heredity,* 24, 633, 1969.

44. Evans, L. T. and Dunstone, R. L., Some physiological aspects of evolution in wheat, *Aust. J. Biol. Sci.,* 23, 725, 1970.

45. Austin, R. B., Morgan, C. L., Ford, M. A., and Bhagwat, S. G., Flag leaf photosynthesis of *Triticum aestivum* and related diploid and tetraploid species, *Ann Bot.,* 49, 177, 1982.

46. Pearlman, I., Thomas, S. M., and Thorne, G. N., Effect of nitrogen fertilizer on photosynthesis of several varieties of spring wheat, *Ann Mot.,* 43, 613, 1979.

47. Kulshrestha, V. P. and Tsunoda, S., The role of "Norin 10" dwarfing genes in photosynthetic and respiratory activity in wheat leaves, *Theor. Appl. Genet.,* 60, 81, 1981.

48. Vishitani, S., Kobayashi, S., and Tsunoda, S., Photosynthesis and growth characteristics of UZU isogenic lines in barley, *Jpn. J. Breed.,* 30(Suppl. 2), 160, 1980.

49. Fakorede, M. A. B. and Mock, J. J., Changes in morphological traits associated with recurrent selection for grain yield in maize, *Euphitica,* 27, 397, 1978.

50. Crosbie, T. M., Pearce, R. B., and Mock, J. J., Selection for high CO_2 exchange rate among inbred lines of maize, *Crop Sci.,* 21, 629, 1981.

51. Curtis, P. E., Ogren, W. L., and Hageman, R. H., Varietal effects in soybean photosynthesis and photorespiration, *Crop. Sci.,* 9, 323, 1969.

52. Harrison, S. A., Boerma, H. R., and Ashley, D. A., Heritability of canopy-apparent photosynthesis and its relationship to seed yield in soybeans, *Crop Sci.,* 21, 222, 1981.

53. Wells, R., Schubze, L. L., Ashley, D. A., Beorma, H. R., and Brown, R. H., Cultivar differences in canopy apparent photosynthesis and their relationship to seed yield in soybeans, *Crop. Sci.,* 22, 886, 1982.

54. Osada, A., Relationship between photosynthetic activity and dry matter production in rice varieties, especially as influenced by nitrogen supply, *Bull. Natl. Inst. Agric. Sci.,* Service D, 14, 117, 1966.

55. Hanson, W. D., Selection for differential productivity among juvenile maize plants, associated with net photosynthetic rate and leaf area changes, *Crop Sci.,* 11, 334, 1971.

56. Berdahl, J. D., Rasmussen, D. C., and Moss, D. N., Effect of leaf area on photosynthetic rate, light penetration and grain yield in barley, *Crop Sci.,* 12, 177, 1972.

57. Eagles, C. G. and Othman, B. D., Regulation of leaf expansion in *Dactyllis,* Report Welsh Plant Breeding Station, Aberystwyth, Wales, 1974, 12.

58. Dunstone, R. L., Gifford, R. M., and Evans, L. T., Photosynthetic characteristics of modern and primitive wheat species in relation to ontogeny and adaptation to light, *Aust. J. Biol. Sci.,* 26, 295, 1973.

59. Bowes, G., Ogren, W. L., and Hageman, R. H., Light saturation, photosynthetic rate, RuBP carboxylase activity and specific leaf weight in soybean grown under different light intensities, *Crop. Sci.,* 12, 77, 1972.

60. Apel, P. and Tschapa, M., Die Akrenatung under phase der Kornfulling bei Weizen *(Triticum aestivum L.),* Biochem. Physiol. Pflanz.,* 164, 266, 1973.

61. Morgan, C. L. and Austin, R. B., Respiratory loss of recently assimilated carbon in wheat, *Ann. Bot.,* 51, 85, 1983.

62. Wilson, D., Variation in leaf respiration in relation to growth and photosynthesis of *Lolium, Ann. Appl. Biol.,* 80, 323, 1975.

63. Robson, M. J., The growth and development of simulated swards of perennial ryegrass. II. Carbon assimilation and respiration in a seedling sward, *Ann. Bot.,* 37, 501, 1973.

64. Heichel, G. H., Confirming measurements of respiration and photosynthesis with dry matter accumulation, *Photosynthetica,* 5, 95, 1971.

65. Penning de Vries, F. W. T., Substrate utilization and respiration in relation to growth and maintenance in higher plants, *Neth. J. Agric. Sci.,* 22, 40, 1974.

66. Wilson, D., Dark respiration, *Report Welsh Plant Breed. Sta. for 1976,* p. 126, 1977.

67. Wilson, D., Plant design and the physiological limitations to production from temperate forages, XIII Int. Bot. Congr. Abstr., Sydney, Australia, 206, 1981.

68. Robson, M. J., The growth and carbon economy of selection lines of *Lolium perenne* CV S23 with differing rates of dark respiration. II. Growth as young plants from seed, *Ann. Bot.,* 49, 331, 1982.

69. Wilson, D., Response to selection for dark respiration rate of mature leaves in *Lolium perenne* and its effects on growth of young plants and simulated swards, *Ann. Bot.,* 49, 303, 1982.

70. Welch, L. F., Nitrogen Use and Behaviour in Crop Production, Bulletin 761, Agricultural Experiment Station, College of Agriculture, University of Illinois, Urbana, 1979, 31.

71. Zieserl, J. F. and Hageman, R. H., Effects of genetic composition on nitrate reductase activity in maize, *Crop Sci.,* 2, 512, 1962.

72. Hageman, R. H., Zieserl, J. F., and Leng, E. R., Levels of nitrate reductase activity in inbred lines and F_1 hybrids in maize, *Nature (London),* 197, 263, 1963.

73. Schrader, L. E., Peterson, D. M., Leng, E. R., and Hageman, R. H., Nitrate reductase activity of maize hybrids and their parental inbreds, *Crop Sci.,* 6, 169, 1966.
74. Croy, L. T. and Hageman, R. H., Relationship of nitrate reductase activity to grain protein production in wheat, *Crop. Sci.,* 210, 280, 1970.
75. Eilrich, G. L. and Hageman, R. H., Nitrate reductase activity and its relationship to accumulation of vegetative and grain nitrogen in wheat *(Triticum aestivum* L.), *Crop. Sci.,* 13, 59, 1973.
76. Warner, R. L., Hageman, R. H., Dudley, J. W., and Lambert, R. J., Inheritance of nitrate reductase activity in *Zea mays* L., *Proc. Natl. Acad. Sci. U.S.A.,* 62, 785, 1969.
77. Zieserl, J. F., Rivenbark, W. L., and Hageman, R. H., Nitrate reductase activity, protein content and yield of four maize hybrids of varying plant populations, *Crop. Sci.,* 3, 27, 1963.
78. Deckard, E. J., Lambert, R. J., and Hageman, R. H., Nitrate reductase activity in corn leaves as related to yield of grain and grain protein, *Crop Sci.,* 13, 343, 1973.
79. Brunetti, N. and Hageman, R. H., Comparison of *in vivo* and *in vitro* assays of nitrate reductase in wheat *(Triticum aestivum* L.) seedlings, *Plant Physiol.,* 58, 583, 1976.
80. Dalling, M. J., Halloran, G. M., and Wilson, J. H., The relation between nitrate reductase activity and grain nitrogen productivity in wheat, *Aust. J. Agric. Res.,* 26, 1, 1975.
81. Shaner, D. L. and Boyer, J. S., Nitrate reductase activity in maize *(Zea mays* L.) leaves. I. Regulation by nitrate flux, *Plant Physiol.,* 58, 499, 1976.
82. Rao, K. P., Rains, D. W., Qualset, C. O., and Huffaker, R. C., Nitrogen nutrition and grain protein in two spring wheat genotypes differing in nitrate reductase activity, *Crop. Sci.,* 17, 283, 1977.
83. Klepper, L. A., Flesher, D., and Hageman, R. H., Generation of reduced nicotinamide dinucleotide for nitrate reduction in green leaves, *Plant Physiol.,* 48, 580, 1971.
84. Hageman, R. H., Reed, A. J., Femmer, R. A., Sherrard, J. H., and Dalling, M. J., Some new aspects of the *in vivo* assay for nitrate reductase in wheat *(Triticum aestivum* L.) leaves, *Plant Physiol.,* 65, 27, 1980.
85. Reed, A. J. and Hageman, R. H., Relationship between nitrate uptake, flux, and reduction and the accumulation of reduced nitrogen in maize *(Zea mays* L.) I. Genotypic variation, *Plant Physiol.,* 66, 1179, 1980.
86. Reed, A. J. and Hageman, R. H., Relationship between nitrate uptake, flux, and reduction and the accumulation of reduced nitrogen in maize *(Zea mays* L.). II. Effect of nutrient nitrate concentration, *Plant Physiol.,* 66, 1184, 1980.
87. Dunand, R. T., Divergent Phenotypic Recurrent Selection for Nitrate Reductase Activity and Correlated Responses in Maize, Ph.D. thesis, University of Illinois, Urbana, 1980.
88. Sherrard, J. H., Lambert, R. J., Messmer, M. J., Below, F. E., and Hageman, R. H., Search for useful physiological and biochemical traits in maize, in *Crop Breeding. A Contemporary Basis,* Vose, P. and Blixt, S. G., Eds., Pergamon Press, New York, 1983, 51.
89. Messmer, M. J., Divergent Phenotypic Recurrent Selection for Nitrate Reductase Activity and Correlated Responses for Various Traits in Corn *(Zea mays* L.), M. S. thesis, University of Illinois, Urbana, 1981.
90. Krantz, B. A. and Chandler, W. V., Effect of heavy application of nitrogen and potash on corn yields, North Carolina Agricultural Experiment Station Bulletin, 366, 1954.
91. Hallauer, A. R. and Miranda, J. B., *Quantitative Genetics in Maize Breeding,* Iowa State University Press, Ames, 1981.
92. Muruli, B. I. and Paulsen, G. M., Improvement of nitrogen use efficiency and its relationship to other traits in maize, *Maydica,* 26, 63, 1981.
93. Messmer, M. J., Leaf Lamina Reduced Nitrogen Concentration and Nitrate Reductase Activity as Criterion for the Identification of Productive Maize Genotypes, Ph.D. thesis, University of Illinois, Urbana, 1983.

Chapter 6

NITROGEN NUTRITION OF GRAIN LEGUMES

Carlos A. Neyra

TABLE OF CONTENTS

I. INTRODUCTION

The Leguminosae constitute the second largest family of the plant kingdom and include several species of agronomical importance. The usefulness of legumes in enhancing the nitrogen fertility of soils has been known for centuries, and they are still the preferred plants in rotations or mixed crop systems because of their capacity to transfer atmospheric nitrogen (N_2) to soils and plants through the process of biological N_2 fixation.[1,2] Legumes are also grown as forage for animals, as oil crops, for direct consumption of the seeds (grain legumes or pulses), or simply as green manure.

Studies on the nitrogen nutrition and metabolism of grain legumes are both complex and fascinating. First, these plants are able to use, simultaneously or complementary, soil or fertilizer nitrogen and N_2 fixed by a symbiotic association between the legume and specific rhizobia.[3-8] Second, nitrate (NO_3) is the most common form of inorganic nitrogen in agricultural soils,[7,9,10] and interactions between plant NO_3 assimilation and biological N_2 fixation can be expected because of the reductive nature of these two processes.[5,11-15] Third, these legumes are exceptionally high in seed protein, particularly globulins, but are deficient in sulfur-containing aminoacids.[16-19] Fourth, establishment of a successful symbioses is made evident by the formation of active nodules. This activity demands the allocation of specific amounts of photosynthate to meet carbon and energy requirements for nodule development and function. Fifth, during periods of active N_2 fixation, glutamine and asparagine are the main forms of nitrogen exported from the nodules to the shoots of temperate legumes (Lupinus, Vicia, Pisum, etc.). Legumes of tropical origin (Glycine, Phaseolus, Vigna, etc.) export from their nodules, primarily the uneides, allantoin, and allantoic acid.[20-24] All of the above and other aspects of the nitrogen nutrition of grain legumes will be discussed in more detail throughout this chapter, together with the presentation of suggestions and alternatives to help improve the efficiency of utilization of nitrogen sources for grain and protein yields. The discussions will focus primarily on the different agronomical, biochemical, and physiological aspects of the nitrogen nutrition of these crops and particular emphasis will be given to soybeans *(Glycine max* L. Merr.) and drybeans *(Phaseolus vulgaris* L.).

II. NITROGEN INPUTS FOR GRAIN YIELDS

Soybean is the most important cash crop among grain legumes. In fact, this crop alone accounts for almost 50% of the total production of grain legumes.[3] The remaining 50% is distributed among peanuts, drybeans, peas, cowpeas, lentils, and other minor legume crops. U.S. farms produce almost 70% of the world soybeans and to a lesser extent they are also important producers of peanuts, drybeans, and peas.

Unitary yields of most grain legumes have remained virtually unchanged or have increased very little over the last quarter of the century. The large increases in productivity observed in recent years for soybeans have been mostly a consequence of increases in production area with a rather modest increase in yields.[3,25] Modern agricultural technologies, like those associated with the green revolution,[25-27] have benefited cereals more than any other crop and, as a result, the amount of cultivated land is about ten times greater for cereals as compared to pulses.[3,25] Of the nonbiological inputs, the use of fertilizer N has probably been the most important single factor responsible for the increase in cereal yields.[3,9] Conversely, pulses do not respond as favorably as cereals to the application of fertilizer N and, in general, fertilizer N is scarcely used in legumes. For instance, corn receives about 40% of the total fertilizer N applied in U.S. farms while soybeans receive only 0.8%.[28] Nonetheless, recent reports have shown that the potential exists and technology is available for increasing N

Table 1
NITROGEN REQUIREMENTS AND EFFICIENCY OF UTILIZATION BY SOYBEANS AND DRYBEANS

Parameter	Soybeans	Drybeans	Corn
Seed protein (%)	40.0	25.0	10.0
Seed N (%)	6.5	4.0	1.6
Farm yields (kg/ha)	2000	2000	5000
Improved yields (kg/ha)[a]	5000	5000	5000
N-harvest index	0.6	0.6	0.6
Seed N (kg/ha)[b]	325	200	80
Plant N (kg/ha)[b]	542	333	133
Calculated Efficiencies (kg/kg)			
Seed yield/plant N[b]	9.2	12.0	37.6
Seed yield/seed N[b]	15.4	25.0	63.0
Dry wt/plant N[c]	30.0(84)	37.0(83)	101.0(82)

Note: Corn values are included for comparative purposes.

[a] Improved yields were normalized across to facilitate comparisons.
[b] Data calculated from the improved yields (5000 kg/ha).
[c] Values calculated from references given in parenthesis.

inputs and enhancing yields of grain legumes.[5,6,29-32] Maximum yield experiments conducted in New Jersey (N.J. Agricultural Experiment Station) have shown yield potentials of about 6000 kg/ha in soybeans[31] and 4000 kg/ha in drybeans.[32] Realization of those maximal yields at the farm level may not happen immediately but will require larger amounts of N assimilated by the plants. For instance, grain yields at or about 5000 kg/ha will demand the incorporation of 542, 333, and 133 kg N per hectare for soybeans, drybeans, and corn, respectively (Table 1). These data also show that grain legumes need three to four times more N for the production of equivalent seed yields as cereals. The high N requirements of grain legumes is undoubtedly related to the high N and protein content of the seeds, a fact that may contribute to limited yields on a weight basis. In addition, the differences between C-3 (soybeans, drybeans, etc.) and C-4 (corn and others) plants in terms of photosynthetic efficiency may also be responsible for the differences in the efficiency of internal N utilization among these crops. Accordingly, corn shows 3 to 4 times greater efficiency of utilization of plant N in terms of seed or dry matter yields as compared to soybeans or drybeans (Table 1). While grain legumes can benefit from the utilization of inorganic N,[4-6] economic considerations and protection of environmental quality would eventually limit the amount of fertilizer N to be used in farming of legume crops. Unfortunately, symbiotic N_2 fixation by the legume-*rhizobrium* association is reportedly inadequate to obtain maximal yields by grain legumes. For instance, soybeans grown under midwestern U.S. conditions have been shown to obtain 25 to 60% of their total N from N_2 fixation with the remainder being derived from soil sources.[4] LaRue and Patterson[33] estimated that the average fixation rate by soybeans in American agriculture is equivalent to 75 kg N per hectare or 50% of the total plant N at an approximate yield of 1500 kg/ha. The contribution of N_2 fixation to plant N in other legumes such as drybeans may range from 5 to 50% of the total plant N.[6,33] In any case, a substantial amount of plant N is derived from soil or fertilizer N sources.[4,6,33] The values reported above are commonly referred to in the literature, but they do not represent the limits of the contribution by N_2 fixation to plant N. The known variability among host-genotypes, bacterial strains, and specific contributions of host X bacteria in terms of efficiency of N_2 fixation provides the basis for more extensive work on the selection of superior bacterial strains and the breeding of legumes for more efficient symbiosis. Improvements in the ability to establish more efficient N_2 fixing symbiosis may reduce the quantity of inorganic N needed for high yields.

The indigenous rhizobia in soils may range from highly efficient symbionts to those capable of nodule formation but unable to fix nitrogen (nod$^+$ fix$^-$ phenotypes). Singleton and Stockinger[34] recently indicated that a proper evaluation of the relative extent to which ineffective nodules can exist on a plant and not significantly fix N_2 may help explain the variability of responses to inoculation with improved strains into soils containing a naturalized population of rhizobia. Competitive ability of the inoculum for nodulation sites on roots of the host plant and development of active nodules are two important characteristics for effective N_2 fixation. To increase competitiveness for nodule sites is probably the most critical problem to the naturally selected or genetically altered bacterial strains. Effective nodulation is recognized by the production of a reasonable number of nodules, normal in size and high in leghemoglobin content surrounding the nitrogenase-containing bacteroids. This leghemoglobin is unique to plants and found exclusively within the nodules.[35,36] It binds to oxygen and controls the flow of oxygen reaching the O_2-sensitive nitrogenase enzyme. In the case of an inefficient symbiosis, plants do not look normal and are usually chlorotic because of N deficiency. Because nodulation and nitrogen fixation are particularly low during the initial vegetative stages and reach a maximum during reproductive growth,[4-6,37-39] plants may also look chlorotic early in the life cycle of the plant. Data presented by Williams et al.[40] have shown that seedling growth in symbiotically grown soybeans was limited primarily by N availability. Their data also showed a failure of CO_2-enrichment to increase significantly total plant N in *Rhizobium*-dependent plants, which indicated that root nodule development and functioning in such plants was not limited by photosynthate production.[40] By contrast, increasing NH_4NO_3 from 0.0 to 8.0 mM produced 251% more dry matter and 287% more Kjeldahl N in plants grown under 320 ppm CO_2. Mature symbiotic legumes, on the other hand, appear to be primarily limited by carbon supply.[3,40] Increasing the CO_2 concentration of the soybean canopy to about 1000 ppm during reproductive growth resulted in an increase of the amount of N fixed from 75 to 425 kg/ha while the amount of nitrogen obtained from the soil was decreased from 220 to 85 kg/ha. In addition, the total N input was increased from 295 to 510 kg/ha.[3] The latter figure is surprisingly close to the 533 kg of N per hectare required for increasing the production of soybeans to 5000 kg/ha as indicated in Table 1. In addition to the effect on N_2 fixation and nodule mass, the CO_2-enrichment increased the duration of nitrogenase activity and extended the period of effective N_2 fixation into the late stages of seed development.[3] The seasonal peak of nitrogenase activity has been associated with the onset of seed development (R-5) followed by a rapid decline of activity approaching zero at physiological maturity.[5,6,37,38] Postflowering applications of fertilizer N have been successful in increasing seed yields of both drybeans[5,32] and soybeans.[29,30] The response to postflowering applications of fertilizer N depends very much on species, cultivar, and level of soil N.[6,32] Low levels of fertilizer N are recommended as "starter" in low N soils to stimulate plant growth and avoid N stress during nodule initiation at the early vegetative stages of the plants.[4,11,41]

According to the preceding discussion, several strategies can be proposed to help increase N inputs for increasing legume yields: (1) enhancement of the nitrogen-fixation efficiency under limiting soil N levels; (2) engineering of better symbiotic legume-*rhizobium* combinations, able to make the best use of the simultaneous presence of atmospheric and soil or fertilizer N sources; (3) development of plant genotypes with a higher efficiency to use soil or fertilizer N sources during vegetative growth without

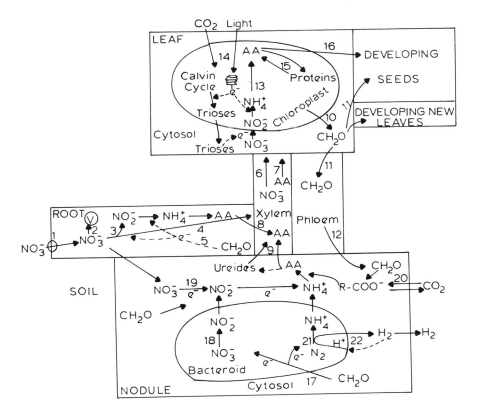

FIGURE 1. Schematic illustration of the different metabolic pathways involved in the assimilation of nitrate and nitrogen fixation in a symbiotic legume. The diagram also illustrates the relationships between nitrate reduction, nitrogen fixation, photosynthate availability, and the mobilization of C and N throughout the plant. The most important features of the diagram are identified by numbers 1 to 22 and each explained within the text.

affecting nodulation and nitrogenase activities; (4) development of plant genotypes capable of responding with higher yields to postflowering N applications; and (5) development of plant genotypes with a longer duration of photosynthetic and N assimilation capabilities well into the seed-development stages.

III. NITROGEN NUTRITION AND METABOLISM IN A SYMBIOTIC LEGUME

As indicated above, grain legumes are able to utilize inorganic soil N (or applied fertilizer) and N_2 fixed by a symbiotic relationship with rhizobia. Both the inorganic and symbiotic N sources are necessary for maximum yields of seed legumes.[4-6] As the soil N or fertilizer N increases, that fixed by the symbiosis decreases.[4,6,11-15] Nitrate is the most common form of inorganic N available to plants in most agricultural soils. Following NO_3 uptake, it can be reduced in roots, stems, and leaves and in symbiotic legumes a fraction of the NO_3 is also reduced within the nodules (Figure 1). Continuous exposure of legume roots to NO_3 affects symbiotic effectiveness by causing a reduction of the number of infected sites (nodule number), nodule development, and the level of nitrogenase activity. The primary cause for this inhibitory effect of NO_3 is yet to be defined and information is still lacking with regard to effective ways to relieve the inhibitory effects of NO_3 on the various aspects of the symbiosis.

A single mechanism of NO_3 inhibition on symbiotic effectiveness is unlikely.[42] First, there is a possibility of energy competition between the processes of NO_3^- reduction and N_2 fixation. The competition for available carbohydrates and reducing equivalents between these two reductive processes can be established within the bacteroids or between the reduction of NO_3^- occurring in different plant tissues and the bacteroid nitrogenase (Figure 1). This competition for energy is also known as the "photosynthate deprivation hypothesis", which attributes the decrease in N_2 fixation activity to a diminished supply of photosynthate to the nodules because of NO_3^- reduction.[13] The second hypothesis invokes a more direct effect and attributes the inhibition of N_2-ase activity by NO_3^- to the accumulation of NO_2^- by the activity of a bacteroid localized nitrate reductase (NR).[14] However, the second mechanism cannot be generalized because nitrate reduction and NO_2^- accumulation does not occur in some legumes containing bacteroids unable to express nitrate reductase activity (NR⁻).[14,43,44] Thus, the sensitivity of the symbiotic system to NO_3^- appears to be a consequence of a combined effect of photosynthate competition along the plant plus the localized effect of NO_2^- accumulation in nodules containing bacteroids expressing NR activity. A third additional hypothesis has been proposed by Berkum and Sloger.[45] These authors working with field-grown soybeans observed that an inverse relationship exists between the concentration of soluble root N and the development of nodule nitrogenase activity. They indicated that the inhibition of N_2 fixation by fertilizer N applications is probably mediated by an increase in the availability of combined N in the root environment.[45] Berkum and Sloger also indicated that a high C/N ratio favors the development of higher levels of nitrogenase activity.[45]

The diagram shown in Figure 1 represents a simplified, integrated view of the different metabolic pathways involved in the utilization of NO_3^- (from soil sources) and atmospheric N_2 by a symbiotic legume. The diagram (Figure 1) also illustrates on the relationships between NO_3^- reduction, N_2 fixation, photosynthate availability, and the mobilization of C and N throughout the plant, including seeds and nodules. The most important features on Figure 1 are identified by numbers as follows. (1) NO_3^- uptake across the plasmalemma of root cells in carrier mediated, requires aerobic metabolism and protein synthesis, exhibits multiphasic kinetics, and is inducible by the substrate.[46-48] The NO_3^- found in root cortex cells can be either stored in vacuoles (2) or reduced to NO_2^-, NH_4^+ and amino acids (3). The remainder of the NO_3^- may move, primarily across cells, to the xylem (4) for its transport to the shoots. The source of reductant for the reduction of NO_3^- to NO_2^- and NO_2^- to NH_4^+ arises from the oxidation of carbohydrates allocated to root cells (5). Nitrate reduction can occur in either roots (2), leaves (13), nodule cytoplasm (19), or within the bacteroids (18). The extent to which reduction and accumulation of incoming NO_3^- occurs in the roots of legumes varies markedly with plant age, species, and the level of NO_3^- currently present in the external solution.[7] Pate and Atkins[7] also indicated that *P. vulgaris, T. repens,* and *G. max* may be classed with *Vigna* as relatively poor reducers of NO_3^- in the roots. Conversely, *Pisum* spp. and *Vicia* spp. joins *Lupinus* spp. as legume species with a high capacity for NO_3^- reduction in their roots.

The fraction of NO_3^- and organic-N reaching to the xylem moves along the transpirational stream, up to the leaves.[6-9] The composition of the xylem sap in relation to N nutrition will depend upon the predominant process (N_2 fixation or NO_3^- assimilation) taking place at certain stages of plant development, the actual level of soil N, and the product of NH_4^+ incorporation. In addition to NO_3^-, amino acids, amides, and ureides (allantoin and allantoic acid) can all be found in xylem sap exudates. During periods of active N_2 fixation, glutamine and asparagine carry up to 90% of the N exported from the nodules to the shoots in temperate legumes (e.g., Lupinus, Pisum, Vicia, etc.). On the other hand, legumes of tropical origin (Glycine, Vigna, Phaseolus, etc.) export primarily the ureides, allantoin, and allantoic acid.[20-24] In ureide-exporting legumes, plants relying on NO_3^- usually export from their roots considerably less ureide and significantly more asparagine compared to N_2-dependent plants. The experimental results obtained have provided strong support for the link between the presence of ureides in the xylem and the occurrence of N_2 fixation in soybeans.[22] Sap from nodulated plants supplied with a N-free nitrution solution contained seasonal averages of 78 and 20% of the total N as ureide-N and amino acid-N, respectively.[49] Other results have suggested that analyses of plant tissue for ureide and NO_3^- (particularly the shoot axis) rather than analyses of xylem sap for ureides and total N may facilitate adaptation of the ureide technique to field studies of nitrogenase activity.[23]

Sucrose is the main form of carbon transported from source leaves towards active sinks, including developing leaves, seeds, roots, and nodules.[11,12] Mobilization of assimilates from source to sink occurs primarily through phloem tissue. Carbon skeletons for sucrose synthesis, in the cytoplasm of mesophyll cells, are exported from the chloroplasts primarily in the form of triosephosphates in exchange for inorganic phosphate (Pi). The exchange process is mediated by a phosphate translocator located on the chloroplast envelope.[50] In the cytoplasm, fructose 6-P and UDP-G react to give sucrose in the presence of sucrose phosphate synthetase.[51] Upon arrival to the demanding sink, sucrose can be rapidly hydrolyzed by the enzyme invertase yielding back glucose and fructose which can be used as respiratory substrates or participate in other biosynthetic reactions. The complete oxidation of sugars helps to provide the NADH, ATP, and carbon intermediates needed for various metabolic reactions.

The relationships between photosynthesis and NO_3^- assimilation has been discussed in a number of recent reports with emphasis on the compartmentation and generation of reductant for NO_3^- and NO_2^- reduction.[52,54] The reduction of NO_3^- (in leaf cells) by a cytoplasmic nitrate reductase is coupled to the oxidation of photosynthetic intermediates and the generation of NADH as electron donor.[52,54] Nitrite reduction occurring in chloroplasts is coupled to photosynthetic electron transport and the generation of reduced ferredoxin as electron donor.[52,55,56] Glutamine and glutamate are primary products of NH_4^+ assimilation in both leaf and nodule cells, as a result of coordinated reactions catalyzed by the GS/GOGAT system.[57] Glutamine and glutamate also play a role in purine and pyrimidine biosynthesis.

The scheme presented in Figure 1 also shows the interactions occurring between the arrival of carbohydrates to the nodules (12) and their utilization to support NO_3^- reduction (18), N_2 fixation (21), and H_2 evolution (22) within the bacteroids. The primary products of NO_3^- reduction (NO_2^-) and N_2 fixation (NH_4^+) are transported out of the bacteroids into the cytoplasm for further metabolism. Any factor affecting the activity of nitrite reductase in the nodule cytosol may lead to the accumulation of toxic levels of NO_2^- and affect nodule function.[14,58] The primary products of NH_4^+ assimilation in the nodule cytosol are again glutamine and glutamate produced by the GS/GOGAT system. Complementary reactions lead to the production of aspartate and asparagine[59,60] or ureides.[7,24]

Nodules of several N_2-fixing plants have an active system of CO_2 fixation (20). Nonphotosynthetic CO_2 fixation mediated by PEP carboxylase have been shown in nodules of several legume species including alfalfa, lupinus, soybeans, and drybeans.[61-64] Approximately 10 to 20% of the photosynthate produced by the plant could be lost through nodule respiration.[65,66] The nodule CO_2 fixation system may then act as a system for recovery of some of the respired CO_2, thus increasing nodule efficiency as well as providing additional carbon skeletons for NH_4^+ assimilation and amino acid biosynthesis.[54,63,64] The efficiency of N_2 fixation is also dependent upon the relative magnitude of H_2 recycling (22). Shubert and Evans[67,68] have shown that some strains of rhizobium possess an uptake hydrogenase (Hup^+ phenotypes) which is membrane bound and participates in the recycling of H_2 and electrons lost during N_2 fixation.

This should result in a more efficient utilization of photosynthate during symbiotic N_2 fixation. H_2 evolution accounts for almost 30% of the energy transferred to the nitrogenase system during the period of active N_2 fixation.[67-69] A recent survey by Lim et al.[69] revealed that more than 75% of the *Rhizobium japonicum* strains present in commercial inocula are Hup$^-$ and the authors concluded that the use of inoculum containing Hup$^+$ strains may help increase soybean yields by enhancing the efficiency of energy utilization by the N_2 fixation process. The latter proposal is based on the assumption that the supply of respiratory substrates to nodules and bacteroids is limiting[3,70] and that more than 25% of the ATP consumption by the reduction of N_2 to NH_4^+ is associated with H_2 evolution.

The reductive assimilation of NO_3^- into amino acids is closely linked to photosynthesis in leaf mesophyll cells (13,14) but the reduction of NO_3^- in roots, nodules, and bacteroids is far removed from the photosynthetic machinery and thus depends upon the allocation of photosynthate to these compartments. Similarly, N_2 fixation is localized within the bacteroids and the link to photosynthesis is established by means of translocation of photosynthetic products.[54] The allocation of carbon substrates to the nodules is needed to meet the energy requirements of N_2 reduction, NH_4^+ assimilation into organic forms as well as for the maintenance of nodule respiratory activity. One of the best demonstrations that net photosynthetic yield is a major limiting factor for N_2 fixation in field grown soybeans came from long-term CO_2 enrichment experiments on the soybean canopy.[3,70,71] These long-term CO_2 enrichment studies were shown to increase nitrogenase activity in soybeans nodules. Conversely, total N_2 fixed was decreased under conditions of elevated pO_2 in the aerial part of soybeans. The major effect of CO_2 enrichment was attributed to an increase in net production of photosynthate by increasing the CO_2/O_2 ratio. The increase in pO_2 increased photorespiration but reduced net photosynthesis and carbohydrate production. On a short-term basis, however, the coupling of photosynthesis and N_2 fixation appears to be rather loose.[72-74] The latter conclusion is supported by a lack of response in terms of nitrogenase activity to CO_2 treatments of up to 36 hr. The authors concluded that nodule activity was not directly limited by current photosynthesis but rather by the partitioning and utilization of photosynthate within the plant. Nodule activity can also be effectively supported by the use of carbohydrate reserves.[72-74]

Sucrose concentration within the nodules has been reported to have a strong influence on acetylene reduction activity of cultured detached nodules of lupins[75] and soybeans.[14] The addition of sucrose (0.1 M) to detached soybean nodules increased nitrogenase activity by about 50% after 2 hr of incubation.[14] These results indicated that soybean nodules were able to take up and metabolize sucrose to forms readily available for the reduction of N_2 to NH_4^+. Also, addition of sucrose (0.1 M) stimulated N_2-ase activity in nodules suspended in a semisolid agar nutrient medium containing NO_3^- or NO_2^-.[14] These results provided support for the contention that increasing the allocation of sugars to the nodules followed by the operation of efficient oxidative systems should provide sufficient reducing equivalents for the simultaneous reduction of N_2, NO_3^-, and NO_2^- metabolism.[14] The question of whether immediately produced photosynthate, carbon reserves, or the partitioning of carbon and competition among active sinks is secondary to the fact that increasing the availability of photosynthate to the nodules enhances N_2 fixation.

IV. CONCLUDING REMARKS

We have provided throughout this chapter supporting evidence for the contention that both NO_3^- assimilation and biological N_2 fixation are needed for maximal yields of grain legumes. A full understanding of the relationships between NO_3^- assimilation and

N_2 fixation may then help to increase the effective input of N for plant growth and grain yields. It is realized, however, that yield is a complex quantity and as such is influenced by the interaction and proper balance of an adequate environment, appropriate agronomical practices, and genetic yield potentials. Therefore, improvement of a single factor or physiological trait may not necessarily result in higher yields. Nonetheless, in the case of cereal crops, N is often cited as "the most important single factor contributing to the yield increases of cereal crops over the last quarter of the century."

The relatively low yields as compared to cereals, as well as the lower efficiency of N utilization by grain legumes are still the main problems to be resolved from the practical standpoint. The extent to which increasing N input and enhancement of the efficiency of N-use can help overcome the limitations on yield output by grain legumes needs to be carefully evaluated. The efforts to improve the overall N nutrition of grain legumes will require a parallel effort to enhance crop photoproductivity. Equally important will be to define very clearly the interrelationships between photosynthesis, partitioning of assimilate, N nutrition, and crop yields.

By the onset of seed development most grain crops would have assimilated a substantial proportion of the final total N found in a mature plant. As a consequence, it is always expected that soil N fertility levels drop gradually as the season progresses. Maintaining adequate levels of soil N through mineralization or by postflowering applications of fertilizer N or magnification of the N input from N_2 fixation during reproductive growth should result in a decrease of flower and pod abortion. Retention of more flowers and pods increases the frequency of successful fertilizations and a higher number of seeds. Consequently, sink size increases. The activity of those sinks is very much dependent upon the activity of the main metabolic processes occurring within the cells. Sink size times sink activity will determine sink strength. Strong sinks demand more photoassimilate and nitrogen to fill their capacity and may result in larger yields only when all of the physiological requirements associated with seed yields are met. A practical evaluation of the interrelationships among all of these processes can be made by determining the harvest index (HI). This value represents the relationship between the allocation of dry matter or N to the reproductive structures in relation to the vegetative structures. In essence, an improvement in the partitioning of assimilates to the grains results in higher HI.[76]

A direct relationship between photosynthesis and yields has always been assumed, but the experimental evidence has not always given strong support to this assumption or the results have been inconclusive. The experience accumulated indicates that the net photosynthetic rate of a selected leaf or leaves does not necessarily correlate positively with seed yields. However, good correlations have been obtained in soybeans between seasonal canopy photosynthesis (SCP) or photosynthetic conversion efficiency (PCE) and grain yields.[77,78] Photosynthetic conversion efficiency is computed by dividing grain yield by the estimated SCP and measures the efficiency of converting photosynthate into yield. In a sense PCE provides another way to express HI but it is more specific in relation to photosynthesis. Similarly, dividing total plant N by the estimated SCP will be interpreted as the efficiency of photosynthate utilization in the assimilation of N regardless of the route of N incorporation.

Alteration of the environment by CO_2 enrichment during reproductive growth did result in a simultaneous increase of both yields and N_2 fixation.[3,70] These experiments revealed that CO_2 fixation and allocation of photosynthate to the nodules enhanced substantially N_2 fixation and the amount of N incorporated into the plant. Unfortunately, similar values of N_2 fixation as those obtained by artificial CO_2 enrichment have not been realized at the farm level as yet. Nevertheless, those results indicate that a vast potential exists for the improvement of N_2 fixation. This will require a development of superior rhizobium strains, improvement of the host plant to support N_2 fixation, and definition of the best host-bacteria combinations.

Maintenance of photosynthesis and N assimilation throughout a substantial portion of the reproductive stages may also be essential for the improvement of grain yields. Early senescence of leaves leads to yellowing, loss of ribulose bisphosphate carboxylase (RuBPCase) activity, and decreased photosynthate availability.[79] During leaf senescence degradation of leaf protein is associated with an increase in proteolytic activity.[80,81] In general, RuBPCase appears to be lost at a faster rate than soluble protein. Inasmuch as RuBPCase is both the key enzyme in CO_2 fixation and the major plant protein, the regulation of RuBPCase breakdown during senescence is very important to the N economy of plants.

Much of the work conducted to date has dealt with the allocation and redistribution of N, based on N balance studies. From that we have learned that some organs are stronger sinks than others and that a sizeable fraction of the N assimilated and stored in vegetative structures may be remobilized to meet the needs of the developing fruits and seeds. The flow of N to the developing reproductive structures is quite large and at maturity the seeds contain anywhere from 50 to 70% of the total plant N, mainly in the form of seed reserve proteins. During the last few years significant progress has been made to understand the biochemical mechanisms responsible for their synthesis and accumulation in the seeds and they actually represent one of the most important systems for exploring the potential of genetic engineering in plants.[19] The storage proteins have no known enzymatic activity but represent a way to store N; they are insoluble in water but soluble in dilute salt solutions. The property of water insolubility allows the reserve proteins to be stable for relatively long periods of time, from seed maturation to seed germination. The bulk of the protein in legume seeds is synthesized during the rapid expansion phase of the cotyledons.[17] At the stage of maximum protein content the globulin fraction may account for about 85 to 90% of the total protein. Globulins are particularly deficient in sulfur-amino acids. Several molecular approaches have been suggested to help improve the balance of amino acids in legume seeds but we have not seen major breakthroughs or satisfactory solutions to this problem, as yet. Improvement of protein quality for animal and human nutrition in all likelihood will require a combined effort between molecular geneticists and traditional breeders as it has been suggested for many of the physiological and biochemical traits associated with carbon (Volume I) and nitrogen (Volume II) nutrition.

ACKNOWLEDGMENTS

The author wishes to thank Sylvia Taylor and Alice Montana for skillful assitance in the preparation of illustrations and typing of the manuscript, respectively. This is a New Jersey Agricultural Experiment Station Publication No. F-01103-03-84. This work was supported by state funds and U.S. Hatch Act funds. Partial support was also provided by a Rutgers Research Council Grant.

REFERENCES

1. Delwiche, C. C., Legumes: past, present and future, *Bioscience,* 28, 565, 1978.
2. Brill, W. J., Agricultural microbiology, *Sci. Am.,* 2, 198, 1981.
3. Hardy, R. W. F. and Havelka, W. D., Nitrogen fixation research: a key to world food?, in *Food: Politics, Economics, Nutrition and Research,* Abelson, P. H., Ed., *Science* (Spec. Compend.), 178, 1975.

4. Harper, J. E., Soil and symbiotic nitrogen requirements for optimum soybean production, *Crop Sci.,* 14, 255, 1974.
5. Franco, A. A., Pereiran, J. C., and Neyra, C. A., Seasonal patterns of nitrate reductase and nitrogenase activities in *Phaseolus vulgaris* L., *Plant Physiol.,* 63, 421, 1979.
6. Westermann, D. T., Kleinkopf, G. E., Porter, L. K., and Legget, G. E., Nitrogen sources for bean seed production, *Agron. J.,* 73, 660, 1981.
7. Pate, J. S. and Atkins, C. A., Nitrogen uptake, transport and utilization, in *Nitrogen Fixation: Legumes, Vol. 3,* Broughton, W. J., Ed., Oxford University Press, New York, 3, 1983, 245.
8. Hensen, R. A. and Heichel, G. H., Partitioning of symbiotically fixed nitrogen in soybeans and alfalfa, *Crop Sci.,* 24, 986, 1984.
9. Hageman, R. H., Integration of nitrogen assimilation in relation to yield, in *Nitrogen Assimilation of Plants,* Hewitt, E. J. and Cuttings, C. V., Eds., Academic Press, New York, 1979, 591.
10. Beevers, L. and Hageman, R. H., Nitrate and nitrite reduction, in *The Biochemistry of Plants, Vol. 5,* Miflin, B. J., Ed., Academic Press, New York, 1980, 115.
11. Gibson, A. H., Limitations to dinitrogen fixation by legumes, in *Proc. 1st Int. Symp. Nitrogen Fixation,* Newton, W. E. and Nyman, C. J., Eds., Washington State University Press, Pullman, 1976, 400.
12. Munns, D. N., Mineral nutrition and legume symbiosis, in *A Treatise on Dinitrogen Fixation, Section IV: Agronomy and Ecology,* Hardy, R. W. F. and Gibson, A. H., Eds., John Wiley & Sons, New York, 1977, 353.
13. Manhart, J. R. and Wong, P. P., Nitrate effect on nitrogen fixation (acetylene reduction): activities of legume root nodules induced by rhizobia with varied nitrate reductase activities, *Plant Physiol.,* 65, 502, 1980.
14. Stephens, B. D. and Neyra, C. A., Nitrate and nitrite reduction in relation to nitrogenase activity in soybean nodules and *Rhizobium japonicum* bacteroids, *Plant Physiol.,* 71, 731, 1983.
15. Timpo, E. E. and Neyra, C. A., Expression of nitrate and nitrite reductase activities under various forms of nitrogen nutrition in *Phaseolus vulgaris* L., *Plant Physiol.,* 72, 71, 1983.
16. Altschul, A. M., Yatsu, L. Y., Ory, R. L., and Engleman, E. M., Seed proteins, *Annu. Rev. Plant Physiol.,* 17, 113, 1966.
17. Millerd, A., Biochemistry of legume seed proteins, *Annu. Rev. Plant Physiol.,* 26, 53, 1975.
18. Chandler, P. M., Higgins, T. J. V., Randall, P. J., and Spencer, D., Regulation of legumin levels in developing pea seeds under conditions of sulfur deficiency: rates of legumin synthesis and levels of legumin mRNA, *Plant Physiol.,* 71, 47, 1983.
19. Larkins, B. A., Genetic engineering of seed storage proteins, in *Genetic Engineering of Plants, An Agricultural Perspective,* Kosuge, T., Meredith, C. P., and Hollander, A., Eds., Plenum Press, New York, 1983, 93.
20. Pate, J. S., Atkins, C. A., White, S. T., Rainbird, R. M., and Woo, K. C., Nitrogen nutrition and xylem transport of nitrogen in ureide-producing grain legumes, *Plant Physiol.,* 65, 961, 1980.
21. Pate, J. S., Transport and partitioning of nitrogenous solutes, *Annu. Rev. Plant Physiol.,* 31, 313, 1980.
22. McClure, P. R., Israel, D. W., and Volk, R. J., Evaluation of the relative ureide content of xylem sap as an indicator of N_2 fixation in soybeans: greenhouse studies, *Plant Physiol.,* 66, 720, 1980.
23. Herridge, D. F., Relative abundance of ureides and nitrate in plant tissues of soybeans as a quantitative assay of nitrogen fixation, *Plant Physiol.,* 70, 1, 1982.
24. Atkins, C. A., Pate, J. S., Ritchie, A., and Peoples, M. B., Metabolism and translocation in ureide-producing grain legumes, *Plant Physiol.,* 70, 476, 1982.
25. Turnham, D., Sources of agricultural growth, in *World Development Report,* Oxford University Press, Oxford, 1982, 57.
26. Borlaug, N. E., Contributions of conventional plant breeding to food production, *Science,* 219, 689, 1983.
27. Simmonds, N. W., Plant breeding: the state of the art, in *Genetic Engineering of Plants. An Agricultural Perspective,* Kosuge, T., Meredith, C. P., and Hollaender, A., Eds., Plenum Press, New York, 1983, 5.
28. Hardy, R. W. F., Filner, P., and Hageman, R. H., Nitrogen input, in *Crop Productivity — Research Imperatives,* Brown, A., Byerly, T. C., Gibbs, M., and San Pietro, A., Eds., Waverly Press, Baltimore, 1975, 133.
29. Garcia, L. R. and Hanway, J. J., Foliar fertilization of soybeans during the seed filling period, *Agron. J.,* 68, 653, 1976.
30. Brevedan, R. E., Egli, D. B., and Legget, J. E., Influence of N nutrition on flower and pod abortion and yield of soybeans, *Agron. J.,* 70, 81, 1978.
31. Flannery, R. L., Influence of soil and crop management factors of soybean yields in New Jersey, in *Proc. 12th Soybean Seed Research Conf.,* American Seed Trade Association, Washington, D.C., 1982, 51.

32. Neyra, C. A. and Pollack, B. L., Seed and Protein Yields of Drybeans Grown in New Jersey, Bulletin No. R-01103-1-82, New Jersey Agricultural Experiment Station, 1982.
33. LaRue, T. A. and Patterson, T. G., How much nitrogen do legumes fix?, *Adv. Agron.*, 34, 15, 1981.
34. Singleton, P. W. and Stockinger, K. R., Compensation against ineffective nodulation in soybean, *Crop Sci.*, 23, 69, 1983.
35. Bergersen, F. J., Leghemoglobin, oxygen supply and nitrogen fixation: studies with soybean nodules, in *Basic Life Sciences, Vol. 10,* Hollander, A., Ed., Plenum Press, New York, 1978, 247.
36. Wittinberg, J. B., Utilization of leghemoglobin-bound oxygen by *Rhizobium* bacteroids, in *Nitrogen Fixation, Vol. 2, Symbiotic Associations and Cyanobacteria,* Newton, W. E. and Orme-Johnson, W. H., Eds., University Park Press, Baltimore, 1980, 53.
37. Harper, J. E. and Hageman, R. H., Canopy and seasonal profiles of nitrate reductase in soybeans *(Glycine max* L. Merr.), *Plant Physiol.*, 49, 146, 1972.
38. Thibodeau, P. S. and Jaworski, E. G., Patterns of nitrogen utilization in soybean, *Planta*, 127, 133, 1975.
39. Nelson, D. R., Bellville, R. J., and Porter, C. A., Role of nitrogen assimilation in seed development of soybean, *Plant Physiol.*, 74, 128, 1984.
40. Williams, L. E., DeJong, T. M., and Phillips, D. A., Carbon and nitrogen limitations on soybean seedling development, *Plant Physiol.*, 68, 1206, 1981.
41. Mahon, J. D. and Child, J. J., Growth response of inoculated peas *(Pisum sativum)* to combined nitrogen, *Can. J. Bot.*, 57, 1678, 1979.
42. Harper, J. E. and Gibson, A. H., Differential nodulation tolerance to nitrate among legume species, *Crop Sci.*, 24, 797, 1984.
43. Manhart, J. R. and Wong, P. P., Nitrate reductase activities and rhizobia and the correlation between nitrate reduction and nitrogen fixation, *Can. J. Microbiol.*, 25, 1169, 1979.
44. de Vasconcelos, L., Miller, L., and Neyra, C. A., Free-living and symbiotic characteristics of chlorate resistant mutants of *Rhizobium japonicum, Can. J. Microbiol.*, 26, 338, 1980.
45. van Berkum, P. and Sloger, C., Interaction of combined nitrogen with the expression of root associated nitrogenase activity in grasses and with the development of N_2 fixation in soybean *(Glycine max* L. Merr), *Plant Physiol.*, 72, 741, 1983.
46. Neyra, C. A. and Hageman, R. H., Nitrate uptake and induction of nitrate reductase in excised corn roots, *Plant Physiol.*, 56, 692, 1975.
47. Israel, D. W. and Jackson, W. A., Ion balance, uptake, and transport processes in N_2-fixing, nitrate and urea-dependent soybean plants, *Plant Physiol.*, 69, 171, 1982.
48. Breteler, H., Hanisch Ten Cate, C. H., and Per Nissen, Time course of nitrate uptake and nitrate reductase activity in nitrogen-depleted dwarf bean, *Physiol. Plant.*, 47, 49, 1979.
49. McClure, P. R. and Israel, D. W., Transport of nitrogen in the xylem of soybean plants, *Plant Physiol.*, 64, 411, 1979.
50. Heldt, H. W., Chon, C. J., Maronde, D., Herold, A., Stankovic, Z. S., Walker, D. A., Kraminer, A., Kirk, M. R., and Heber, W., Role of orthophosphate and other factors in the regulation of starch formation in leaves and isolated chloroplasts, *Plant Physiol.*, 59, 1146, 1983.
51. Giaquinta, R. T., Phloem loading of sucrose, *Annu. Rev. Plant Physiol.*, 34, 347, 1983.
52. Neyra, C. A. and Hageman, R. H., Dependence of nitrite reduction on electron transport in chloroplasts, *Plant Physiol.*, 54, 480, 1974.
53. Neyra, C. A. and Hageman, R. H., Pathway for nitrate assimilation in corn *(Zea mays* L.) leaves: cellular distribution of enzymes and energy sources for nitrate reduction, *Plant Physiol.*, 62, 618, 1978.
54. Neyra, C. A., Interactions of plant photosynthesis with dinitrogen fixation and nitrate assimilation, in *Limitations and Potentials of Nitrogen Fixation in the Tropics,* Dobereiner, J., Burris, R., and Hollander, A., Eds., Basic Life Sciences, Vol. 10, Plenum Press, New York, 1978, 111.
55. Miflin, B. J., The location of nitrite reductase and other enzymes related to amino acid biosynthesis in the plastids of roots and leaves, *Plant Physiol.*, 54, 550, 1974.
56. Magalhaes, A. C., Neyra, C. A., and Hageman, R. H., Nitrite assimilation and amino nitrogen synthesis in isolated spinach chloroplasts, *Plant Physiol.*, 53, 411, 1974.
57. Miflin, B. J. and Lea, P. J., Amino acid metabolism, *Annu. Rev. Plant Physiol.*, 28., 299, 1977.
58. Stretter, J. G., Synthesis and accumulation of nitrite in soybean nodules supplied with nitrate, *Plant Physiol.*, 69, 1429, 1982.
59. Bolland, M. J., Farnden, K. J. F., and Robertson, J. G., Ammonia assimilation in nitrogen-fixing legume nodules, in *Nitrogen Fixation,* Vol. 2, Newton, W. E. and Orme-Johnson, W. H., Eds., University Park Press, Baltimore, 1980, 33.
60. Scott, D. B., Farnden, K. J. F., and Robertson, J. G., Ammonia assimilation in lupin nodules, *Nature (London)*, 263, 703, 1976.
61. Lawrie, A. C. and Wheeler, C. J., Nitrogen fixation in the root nodules of *Vicia fava* L. in relation to the assimilation of carbon. II. The dark fixation of carbon dioxide, *New Phytol.*, 74, 437, 1975.

62. Christeller, J. T., Laing, W. A., and Sutton, W. D., Carbon dioxide fixation by lupin root nodules. I. Characterization, association with phosphoenol pyruvate carboxylase, and correlation with nitrogen fixation during nodule development, *Plant Physiol.*, 60, 47, 1977.

63. Coker, Q. T., III and Schubert, K. R., Carbon dioxide fixation in soybean root and nodules. I. Characterization and comparison with N_2 fixation and composition of xylem exudate during early nodule development, *Plant Physiol.*, 67, 691, 1981.

64. Vance, C. P., Stade, S., and Maxwell, C. A., Alfalfa root nodule carbon dioxide fixation. I. Association with nitrogen fixation and incorporation into amino acids, *Plant Physiol.*, 72, 469, 1983.

65. Minchin, F. R. and Pate, J. S., The carbon balance of a legume and the functional economy of its root nodules, *J. Exp. Bot.*, 24, 295, 1983.

66. Layzell, D. B., Rainbird, R. M., Atkins, C. A., and Pate, J. S., Economy of photosynthate use in N-fixing legume nodules: observations on two contrasting symbioses, *Plant Physiol.*, 64, 888, 1979.

67. Schubert, K. R. and Evans, H. J., Hydrogen evolution: a major factor affecting the efficiency of nitrogen fixation in nodulated symbionts, *Proc. Natl. Acad. Sci. U.S.A.*, 73, 1207, 1976.

68. Schubert, K. R., Engelke, J. A., Russell, S. A., and Evans, H. J., Hydrogen reactions of nodulated leguminous plants. I. Effect of rhizobial strain and plant age, *Plant Physiol.*, 60, 651, 1977.

69. Lim, S. T., Uratsu, S. L., Weber, D. F., and Keyser, H. H., Hydrogen uptake (hydrogenase) activity of *Rhizobium japonicum* strains forming nodules in soybean production areas of the U.S.A., in *Genetic Engineering of Symbiotic Nitrogen Fixation and Conservation of Fixed Nitrogen,* Lyons, J. M., Valentine, R. C., Phillips, D. A., Rains, D. W., and Huffaker, R. C., Eds., *Basic Life Sciences,* Vol. 17, Plenum Press, New York, 1981, 159.

70. Havelka, U. D. and Hardy, R. W. F., Legume nitrogen fixation as a problem in carbon nutrition, in *Proc. 1st Int. Symp. Nitrogen Fixation, Vol. 2,* Newton, W. E. and Nyman, C. J., Eds., Washington State University Press, Pullman, Washington, 1976, 456.

71. Quebedaux, B., Havelka, W. D., Livak, K. L., and Hardy, R. W. F., Effects of altered pO_2 in the aerial part of soybeans on symbiotic N_2 fixation, *Plant Physiol.*, 56, 761, 1975.

72. Schweitzer, L. E. and Harper, J. E., Effect of light, dark and temperature on root nodule activity (acetylene reduction) of soybeans, *Plant Physiol.*, 65, 51, 1980.

73. Finn, G. A. and Brun, W. A., Effect of atmospheric CO_2 enrichment on growth, nonstructural carbohydrate content and root nodule activity in soybean, *Plant Physiol.*, 69, 327, 1982.

74. Heichel, G. H. and Vance, C. P., Physiology and morphology of perennial legumes, in *Nitrogen Fixation: Legumes, Vol. 3,* Broughton, W. J., Ed., Oxford University Press, New York, 1983, 91.

75. Laing, W. A., Christeller, J. T., and Sutton, W. D., Carbon dioxide fixation by lupin nodules. II. Studies with ^{14}C-labeled glucose, the pathway of glucose catabolism and the effects of some treatments that inhibit nitrogen fixation, *Plant Physiol.*, 63, 450, 1979.

76. Gilford, R. M., Thorne, J. H., Hitz, W. D., and Giaquinta, R. T., Crop productivity and photoassimilate partitioning, *Science*, 225, 801, 1984.

77. Christy, A. L. and Porter, C. A., Canopy photosynthesis and yield in soybeans, in *Photosynthesis, Vol. 2, Development, Carbon Metabolism, and Plant Productivity,* Govindjee, Ed., Academic Press, New York, 1982, 499.

78. Wells, R., Schulze, L. L., Ashley, D. A., Boerma, H. R., and Brown, R. H., Cultivar differences in canopy apparent photosynthesis and their relationship to seed yield in soybeans, *Crop Sci.*, 22, 886, 1982.

79. Wittenbach, V. A., Ribulose bisphosphate carboxylase and proteolytic activity in wheat leaves from anthesis through senescence, *Plant Physiol.*, 64, 884, 1979.

80. Wittenbach, V. A., Ackerson, R. C., Giaquinta, R. T., and Herbert, R. R., Changes in photosynthesis, ribulose bisphosphate carboxylase, proteolytic activity and ultrastructure of soybean leaves during senescence, *Crop Sci.*, 20, 225, 1980.

81. Wittenbach, V. A., Breakdown of ribulose bisphosphate carboxylase and change in proteolytic activity during dark-induced senescence of wheat seedlings, *Plant Physiol.*, 62, 604, 1978.

82. De-Polli, H., Nitrogenase Activity in Roots and Stems of Corn *(Zea mays* L.) Cultivars and Its Reponse to Nitrogen Fertilizer and Lime Amendments, Ph.D. thesis, Rutgers University, New Brunswick, N.J., 1983.

83. Timpo, E. E., Nitrate Uptake, Assimilation and the Expression of Nitrate Reductase Activities Under Various Forms of Nitrogen Nutrition in *Phaseolus vulgaris* L., Ph.D. thesis, Rutgers University, New Brunswick, N.J., 1983.

84. Stephens, B. D., Nitrate and Nitrite Reduction in Relation to Nitrogenase Activity in Soybean Nodules and *Rhizobium japonicum* Bacteroids, Ph.D. thesis, Rutgers University, New Brunswick, N.J., 1985.

INDEX

of legumes, 132—134, 138—140
nitrate reductase activity in maize and, 117, 121, 123, 125—126
Growth regulators, 85, 87, 89

H

Half-sib family selection method, 121—123
Harvest index (HI), 139
Haworthia, 20
HD formation, 46
Heavy metals, 5, 11, 23, 25
H^+-efflux pump, 82—83, 87
Heritability, 110, 113, 117, 121—122
Higher plants
genomes in, 60
iron assimilation pathway of, 11
nitrate reductase of, 3—4, 6—8, 12—14, 25
nitrogen sources in, 19
Holo-nitrate reductase, 6—8, 22—23, 26
Host-plant specificity, 53
Hybridization, homologous, 63
Hydrogenase, 45, 137
Hydroxylamine, 11
Hydroxyl efflux, 83—85
p-Hydroxymercuribenzoate, 11

I

Immunochemistry, 15—17
Initiation site, 60
Introns, 60, 71
Iron assimilation pathway, 11
Iron chelates, 10—11, 25
Iron-citrate reductase, 11
γ-Irradiation, 20

K

Kernel growth, cessation of, 111
Kinetin, 89
Klebsiella pneumoniae, 47—50

L

Leaf cells, 93—95
Leaf veins, 95
Leaf water potential, 112
Leaves
ammonium influence on, 19
nitrate reductase
activity in, 19, 117—119
stabilization in, 5—6
nitrate reduction in, 4
senescence in, 140
Lectin, 54, 69
Leghemoglobin, 53, 134

Legumes, 52—55, 132—140
Legumin, 62
Lemna, 71, 80, 83—84, 89, 95
gibba, 83
paucicostata, 78, 80
Leupeptin, 7
Light, 19, 94
Lolium perenne, see also Ryegrass, 113—114
Lupinus spp., 136—138

M

Maize, see also Corn, 109—127
mitochondrial genome of, 68—69
nitrate
reduction in, 90
uptake by, 78—79, 82, 84—85, 95
nitrate reductase activity in, 22, 94, 115—121
nitrogen metabolism as a selection trait
nitrate reductase, 115—121
nitrogen content or concentration, 121—127
nitrogen supply, response to, 74—75
nitrogen use efficiency in, 97
photosynthetic activity as a selection trait, 113—114
plastid genome of, 67
selection traits in, 110—127
yield
factors limiting, 111—112
of hybrid, 110
nitrate reductase activity and, 117, 121, 123, 125—126
nitrogen content and, 121—127
Malate, 87
Malate/aspartate shuttle, 25
Male fertility, 68—69
Mass-selection breeding methods, 121—123
Membrane energization, 51
Membrane potential, 83—84, 86, 88
Mersalyl, 84
Mesophyll cells, 4, 20
Methionine sulfoximine, 48
Methylene blue, 52
Methyl isocyanide, 46—47
6-Methylpurine, 91
Methyl viologen, 10, 23
MgADP, 51
MgATP, 43—45, 51—52
Michaelis constant, 45
Mineral nutrition, 19
Mitochondria, 11, 19, 25, 68
Mitochondrion-coded proteins, 60, 68—71
Mo-cofactor, 2, 7—8, 15—16, 20—21, 26
MoFe protein, 42, 49
Molybdate, 22
Molybdenum
deficiency, apo-NR in plants and cells with, 15, 20
nitrate reductase activity and, 26
nitrogenase protein synthesis and, 50

O

O$_2$, 43, 47, 49
Operon model, 47
Organic acids, 19, 80—83

P

Pea, 26
  bud tips, 19
  plastid genome of, 67—68
Penicillium chrysogenum, 6, 13
PEP carboxylase, 137
Pest resistance, 121
pH, 88
Phaseolin, 70
Phaseolus, 64, 67
  vulgaris, 14, 132, 136
Phosphate uptake, 87, 137
Photorespiration, 19—20
Photosynthate deprivation hypothesis, 136
Photosynthesis
  energy from, 53
  genetic variability for, 110
  nitrate absorption and 86—87, 137
  yield and, 139
Photosynthetic activity, 113—114, 126, 134
Photosynthetic conversion efficiency (PCE), 139
Phytochrome, 19
Phytohormones, 14, 19—20
Phytoplankton, 85
Ping-pong mechanism, 11—12, 14, 18
Pink Sepharose, 5, 8, 28
Pisum sativum, see Pea
Plant breeding, conventional, 110
Plastid-coded proteins, 60, 64—68
Polycistronic messages, 66
Posttranscriptional processing, 69—71
Posttranslational modifications, 61—62, 65
Posttranslational processing, 62, 69
Potassium, 81—82, 87, 91
ppGpp, 50
Pribnow box, 60, 69—70
Promoter, 60, 69—71
Proplastids, 64, 71
Proteinases, 7, 24, 26
Protein synthesis, 19, 59—71
  mitochondria-coded proteins, 68—69
  nuclear-coded proteins, 60—66
  plastid-coded proteins, 68—69
  regulation of gene expression in plants, 69—71
Protein synthesis inhibitors, 20, 22, 24, 84
Protons, 45
Pseudomonas fluorescens, 80
Pulses, 132
Purple nonsulfur photosynthetic bacteria, 51—52

Q

Quinones, 10

R

Radish cotyledons, 18
Regulatory genes, 21
Respiration rate, as selection trait, 114
Restriction enzyme, 63
Rhizobia, 52—54, 134, 137
Rhizobium japonicum, 45, 138
Rhizobium:legume symbiosis, 54, 133—134
Rhodopseudomonas
  palustris, 50
  sphaeroides, 51
Rhodospirillum rubrum, 51—52
Rhodotorula glutinis, 6, 13
Ribosomes, 61—62
Ribulosebisphosphate carboxylase (RuBPCase), 64—66, 69, 114, 140
Rice
  nitrate reductase in, 14, 22—23
  nitrate uptake by, 87, 89
  photosynthetic rates in, 113—114
  yield, 110—111, 115
RNA polymerase, 60
RNA synthesis, 19, 64—66
  inhibitors, 20, 22, 84, 91
Rocket immunoelectrophoresis, 15, 21
Root length densities, 76
Roots
  ammonium influence on, 19
  cultured, as model system, 19
  nitrate reductase activity in, 19
  nitrate
    reduction in, 4
    uptake in, 74—93
  proliferation of, 76
Rose cell cultures, 20
Ryegrass, see also *Lolium*, 77, 114

S

Scenedesmus, 87
Scutella, 4
Seasonal canopy photosynthesis (SCP), 139
Seed proteins, 62—65, 69—70, 132, 140
Selection traits
  identification of, 112
  nitrate reductase activity as, 115—121
  nitrogen content as, 121—127
  photosynthetic activity as, 113—114
  physiological and biochemical, 110, 122
  requirements for, 110
Senescence, leaf, 140
Signal peptide, 62

3 5282 00237 9462